Lecture Notes in Mathematics

Edited by A. Dold and B. Eckmann

552

C. G. Gibson K. Wirthmüller
A. A. du Plessis E. J. N. Looijenga

Topological Stability of Smooth Mappings

Springer-Verlag
Berlin · Heidelberg · New York 1976

Authors

Christopher G. Gibson
Department of Pure Mathematics
University of Liverpool
P. O. Box 147
Liverpool, L69 3BX/Great Britain

Klaus Wirthmüller
Fachbereich Mathematik d. Universität
Postfach 397
8400 Regensburg/BRD

Andrew A. du Plessis
School of Mathematics and Computer Science
University College of North Wales
Bangor
Gwynedd, LL57 2 UW, Wales/Great Britain

Eduard J. N. Looijenga
Mathematisch Instituut der
Katholieke Universiteit
Toernooiveld
Nijmegen/The Netherlands

AMS Subject Classifications (1970): 57D45, 58C25

ISBN 3-540-07997-1 Springer-Verlag Berlin · Heidelberg · New York
ISBN 0-387-07997-1 Springer-Verlag New York · Heidelberg · Berlin

Printed in Germany

Printing and binding: Beltz Offsetdruck, Hemsbach/Bergstr.

Preface

During the academic year 1974 - 75 the Department of Pure Mathematics in the University of Liverpool held a seminar on the Topological Stability of Smooth Mappings: the main objective was to piece together a complete proof of the Topological Stability Theorem (conjectured already by Thom in 1960, and proved by Mather around 1970) for which no published account existed. This volume comprises a write-up of the seminar by four of its participants.

There are several acknowledgements which should be made. Any mathematician working in this area is conscious of his debt to the inventiveness of Thom, and to the technical work of Mather which has placed much that was conjecture on a firm mathematical foundation. As far as the seminar is concerned I would like to single out the special contribution of Eduard Looijenga, who showed us how to fill in gaps which otherwise might have remained open. Also, I would like to acknowledge the considerable help offered by Terry Wall, from the inception of the seminar to the production of the typescript. Further acknowledgements are due to the seminar audience (who frequently had good reason to appear confused) for their patience and in particular to Tim Ward who helped out with the talks; to the British Scientific Research Council who provided financial support for Eduard Looijenga, and the University of Wales whose financial assistance enabled Andrew du Plessis to participate in the seminar; to Les Lander who was inveigled into drawing the diagrams, and did an excellent job; to Dirk Siersma and Klaus Lamotke whose careful reading of parts of the manuscript removed many errors; and finally to Evelyn Quayle, Jean Owen and Margaret Walker who produced a first-class typescript.

Liverpool, July 1976. C. G. Gibson.

Contents

Introduction

Motivation and some historical remarks.

The C^∞-mappings f and f' from a manifold N to a manifold P are said to be C^ℓ-equivalent ($\ell = 0, 1, 2, \ldots, \infty$) if there exist C^ℓ-automorphisms h of N and h' of P such that $f' = h' \circ f \circ h^{-1}$. This is clearly an equivalence relation: it simply says that f and f' have the same 'C^ℓ-behaviour'. We call a C^∞-mapping C^ℓ-stable if the equivalence class of f (with respect to C^ℓ-equivalence) forms a neighbourhood of f in the function space $C^\infty(N, P)$. This presupposes a topology on $C^\infty(N, P)$: we choose the Whitney topology (see Ch. IV for the definition). Obviously, this property becomes stronger as ℓ increases. There is an important question, related to this notion, which has also some physical interest. Namely, whether any proper C^∞-mapping can be approximated by a C^ℓ-mapping or more precisely, whether among the set of proper C^∞ mappings, $C^\infty_{pr}(N, P)$, the C^ℓ-stable ones are dense. The Morse lemma and the Whitney embedding theorem imply that the answer is yes if $\dim P = 1$ or greater than $2 \dim N$ (and ℓ arbitrary). H. Whitney seems to have been the first to investigate this question in its own right. In 1955 [W1] he showed that the answer is affirmative in case $\dim N = \dim P = 2$. His proof included a fairly simple characterisation of C^∞-stable mappings between surfaces. A few years later, in 1959, Thom [TL] gave examples of proper smooth mappings from $\mathbb{R}^n$ to $\mathbb{R}^n$ $(n \geq 9)$ which cannot be approximated by a C^2-stable mapping. A complete answer to this question for $\ell = \infty$ was given by J.N. Mather around 1967. In his fundamental series 'Stability of C^∞-mappings' [I-VI] he obtains among other things a (multi)-transversality criterion for C^∞-stability. For certain pairs $(n, p) := (\dim N, \dim P)$, this transversality criterion involves uncountably many transversality conditions. As one may expect, in such a case these transversality conditions cannot be simultaneously satisfied by a dense subset of $C^\infty_{pr}(N, P)$. Thus Mather was able to determine the pairs (n, p) for which the answer to our question (with $\ell = \infty$) is yes: the C^∞-stable proper mappings are dense in $C^\infty_{pr}(N, P)$ if and only if (n, p) satisfies one of the following conditions

$$n < \frac{6}{7}p + \frac{8}{7} \quad \text{and} \quad p - n \geq 4 ,$$

$$n < \frac{6}{7}p + \frac{9}{7} \quad \text{and} \quad 3 \geq p - n \geq 0 ,$$

$$p < 8 \quad \text{and} \quad p - n = -1 ,$$

$$p < 6 \quad \text{and} \quad p - n = -2 ,$$

$$p < 7 \quad \text{and} \quad p - n = -3 .$$

Mather also showed that for $\ell = 1$ the answer to our question is in general negative. So we are lead to consider the case $\ell = 0$ (topological equivalence).

In the previously mentioned article [TL], Thom conjectured that the answer would be yes in that case. In later papers [T1, 2] he gave some indications as to how this might be proved. There he introduced an intermediate between the smooth and piecewise-linear categories : the category of stratified spaces and stratified mappings. Part of its interest comes from Whitney's regularity lemma [W2] which asserts that any complex-analytic set admits a stratification in Thom's sense. S. Lojasiewicz [L] generalised this later to semi-algebraic sets and used this extension to prove that any closed semi-algebraic set is triangulable. Thom's work on stratified sets culminated in his famous isotopy lemma's [T1, T3, M1] to which we shall return below. By fully exploiting his own work on C^∞-stability and Thom's philosophy, Mather obtained around 1970 a definite proof of Thom's conjecture. An outline of this proof has been given in Séminaire Bourbaki by Chenciner [C] (see also [M2]). A complete version will appear in a book Mather is writing.

The proof presented in these notes is somewhat different from Mather's (as it does not use the existence of global universal unfoldings) and follows Thom's indications more closely. As far as prerequisites are concerned, we require of the reader no more than some familiarity with differential topology and commutative algebra, both on an elementary level. For a few not-so-standard facts we refer to other work, however. These are various properties of semi-algebraic sets (I-(2.1), (2.2), (2.4), (2.5) and (2.6)) and the Malgrange preparation theorem.

Structure of the proof.

Actually the proof will yield a stronger result. In order to state it, let us call $f \in C^{\infty}(N, P)$ <u>strongly topologically stable</u> if it has a neighbourhood U such that any $f' \in U$ can be joined in U with f by means of a smooth one-parameter family $\{f_t\}_{t \in I}$ (where I stands for the unit interval) which is <u>topologically trivial</u>. By the latter we mean that there exist continuous families of topological automorphisms $\{h_t\}_{t \in I}$ of N and $\{h'_t\}_{t \in I}$ of P such that $f_0 = h'_t \circ f_t \circ h_t^{-1}$ for all $t \in I$. Strong topological stability clearly deserves its name: it implies C^0-stability. Our aim is to prove that the strongly topologically stable mappings are dense in $C^{\infty}_{pr}(N, P)$. For simplicity we take N compact here.

At some stage we have to construct one-parameter families of homeomorphisms. Only one geometric technique is known that can furnish these, namely integration of vector-fields. As is well-known, a C^1-vector-field ξ on a compact manifold M generates a one-parameter family of C^1-diffeomorphisms. In certain situations this is mimicked by possibly discontinuous vector-fields which satisfy certain control conditions. These will then generate a one-parameter family of homeomorphisms. A typical and very illuminating example of such a vector-field is described in (II, 1.1). We advise the reader to have a look at it before going further. Roughly speaking in Thom's stratification theory one considers the case where the vector-field ξ on the manifold M is such that M can be broken up into smooth submanifolds $\{M_i\}$ such that $\xi | M_i$ is smooth and tangent to M_i. In order to be more precise, we need the notion of Whitney regularity. Let X and Y be disjoint submanifolds of a euclidean space. Then Y is called Whitney regular over X if for any pair of sequences $\{x_i \in X\}_{i=1}^{\infty}$, $\{y_i \in Y\}_{i=1}^{\infty}$ both converging to some $x_0 \in X$ such that the lines $\overline{x_i y_i}$ converge to a line ℓ and the tangent planes Ty_iY converge to a plane p, we have $\ell \subseteq p$. This notion is coordinate invariant and has therefore still a meaning if X and Y are in a manifold M instead of euclidean space. The interest of this notion is explained by the following property. If Y is Whitney regular over X, then

X admits a (tubular) neighbourhood T_X in M with retraction $\pi_X : T_X \to X$ such that $\pi_X : (T_X, T_X \cap \bar{Y}, X) \to X$ is a topologically locally trivial fibre bundle triple with 'cone-like' fibres. The proof of this (which is not hard) consists in showing that any ordinary vector-field ξ on X extends to a controlled (possibly discontinuous, but at any rate integrable) vector-field η on T_X such that $\eta|Y$ is tangent along Y and $d\pi_X \circ \eta = \xi \circ \pi_X$. This generalises as follows. Let A be a subset of M which is endowed with a locally finite partition $\mathcal{A}$ into submanifolds of M such that for any pair $X, Y \in \mathcal{A}$, Y is Whitney regular over X. Such a partition is usually called a Whitney stratification and its members strata.

Now suppose that $p : M \to M'$ is a proper submersion such that the restriction of p to any $X \in \mathcal{A}$ is also a submersion. Then Thom's first isotopy lemma asserts that $p : (M, S, \{X \in \mathcal{A}\}) \to M'$ is a locally trivial fibre bundle. Especially interesting is the case where $M = N \times \mathbb{R}$ and $p : N \times \mathbb{R} \to \mathbb{R}$ ordinary projection; then it follows that $A_t = \{x \in N : (x, t) \in A\}$ and A_o are isotopic in N (hence the name isotopy lemma). We will rather be concerned with the second isotopy lemma which deals with families of mappings (instead of families of spaces). Let $F : M \to M'$ be a smooth mapping. A Thom stratification for F is a pair of Whitney stratifications $\mathcal{A}$ of M and $\mathcal{A}'$ of M' such that F maps any stratum of $\mathcal{A}$ submersively into a stratum of $\mathcal{A}'$, and satisfies another condition (Thom regularity) which we do not describe here but which is 'almost always' fulfilled. The case that interests us is when $M = N \times \mathbb{R}$, $M' = P \times \mathbb{R}$ and $F : N \times \mathbb{R} \to P \times \mathbb{R}$ is of the form $(F(x, t) = f_t(x), t)$. In this situation the second isotopy lemma tells us that if F admits a Thom stratification $(\mathcal{A}, \mathcal{A}')$ such that the ordinary projection $P \times \mathbb{R} \to \mathbb{R}$ maps the strata of $\mathcal{A}'$ submersively to $\mathbb{R}$, then $\{f_t\}_{t \in \mathbb{R}}$ is a topologically trivial family. Note that then the strata of $\mathcal{A}$ and $\mathcal{A}'$ must be transverse to $N \times \{t\}$ and $P \times \{t\}$. It is not hard to show that by taking intersections $\mathcal{A}$ and $\mathcal{A}'$ determine a Thom stratification $(\mathcal{A}_t, \mathcal{A}'_t)$ for f_t. We therefore may paraphrase the second isotopy lemma by saying that the family $\{f_t\}$ is topologically trivial if it admits a smooth family of Thom stratifications

$\{(A_t, A'_t)\}$.

In view of this, it seems to be important to construct Thom stratifications for an open-dense set of mappings. This is best done by means of a certain Whitney stratification of the jet space $J^\ell(N, P)$. If A is a locally finite partition of $J^\ell(N, P)$ into manifolds and $f : N \to P$ such that its ℓ-jet extension $J^\ell f : N \to J^\ell(N, P)$ is transverse to the members of A, then $A_o = (J^\ell f)^{-1} A$ is a finite manifold partition of N. We call f <u>multi-transverse with respect</u> to A if moreover f maps the members of A_o in a regular way to P, more precisely if for any $y \in P$ the canonical homomorphism

$$T_y P \to \bigoplus_{x \in f^{-1}(y)} (T_y P / Tf(T_x X_x))$$

(with $X_x \in A_o$ containing x) is a surjection. It can be shown (in much the same way as the Thom transversality lemma is proved) that the mappings which are multi-transverse with respect to A are dense in $C^\infty(N, P)$. If f is multi-transverse with respect to A, then A_o can be refined in a natural way to a manifold partition A_f (namely, the set of indecomposables of the Boolean algebra generated by A_o and $\{f^{-1}f(X) : X \in A_o\}$) such that two members of A_f have either disjoint or equal f-images. We then obtain a partition of $f(N)$ (not necessarily into manifolds, of course). By adding $P \setminus f(N)$ we obtain a partition of P which we denote by A'_f.

In the last chapter it is shown that for ℓ sufficiently large there exists a Whitney stratification $A^\ell(N, P)$ of $J^\ell(N, P)$ with the property that for any $f \in C^\infty(N, P)$ multi-transverse with respect to $A^\ell(N, P)$ the corresponding pair (A_f, A'_f) actually defines a Thom stratification of f. Moreover the mappings multi-transverse with respect to $A^\ell(N, P)$ are not only dense, but also open in $C^\infty(N, P)$. Now let f be multi-transverse with respect to $A^\ell(N, P)$. By local arcwise connectedness it has a neighbourhood U such that any $f' \in U$ can be joined with f by a smooth family $\{f_t\}$ of mappings which are multitransverse with respect to $A^\ell(N, P)$. Then the corresponding Thom stratifications (N, A_{f_t}), (P, A'_{f_t}) fit together to form a Thom stratification (A, A') of the associated $F : N \times I \to P \times I$ with the property that strata of $\mathcal{A}'$ map submersively to I. By the second isotopy lemma f and f' are then topologically equivalent.

We shall not discuss the construction of $\mathcal{A}^{\ell}(N, P)$ here, but only remark that for its definition a very basic rôle is played by the concept of a canonical stratification. In chapter I this notion is investigated in its relation to the semi-algebraic category. The next chapter states and proves the isotopy lemmas of Thom. Chapter III is an exposition of the Thom-Mather unfolding theory and thus contains somewhat more than we need. It is completely independent of the two previous chapters. As we have said, in the last chapter the Whitney stratification $\mathcal{A}^{\ell}(N, P)$ is constructed and by means of this stratification the proof of the stability theorem is completed.

Related problems.

Finally, we mention two related problems, which have already been discussed by Thom in [T2] and [T4].

Problem 1 Has any jet $z \in J^{\ell}(N, P)$ an extension $z' \in J^{\ell'}(N, P)$ $(\ell' \geq \ell)$ which is C^0-sufficient (i.e. such that any two representatives of z' are topologically equivalent)?

This problem has been solved affirmatively by A.N. Varchenko [V], who obtained in fact a much stronger result. Apparently still open is

Problem 2 Does there exist a Whitney stratification $\mathcal{A}$ of $C^{\infty}(N, P)$ (neglecting a subset of infinite codimension) into submanifolds of finite codimension with the following property: a smooth family $\{f_t \in C^{\infty}(N, P)\}$ (with U a parameter manifold) is C^0-stable (in an obvious sense) if the associated map $U \to C^{\infty}(N, P)$ is transverse to $\mathcal{A}$? This is known to be so in the case $\dim P = 1$.

Nijmegen, May 1976 E.J.N. Looijenga

References cited in the introduction.

[C] A. Chenciner, Travaux de Thom et Mather sur la stabilité topologique. Sém. Bourbaki, Février 1973, no. 424. (Springer Lecture Note 383)

[Ł] S. Łojasiewicz, Ensembles semi-analytiques, available at IHES (1965).

[MI-VI] Stability of C^∞-mappings I-VI, Ann. of Math. 87 p.89-104, Ann. of Math. 89 p.254-291, Publ. Math. IHES 35 p.127-156, Publ. Math. IHES 37 p.223-248, Adv. in Math. 4 p.301-366, Proc. Liverpool Symp. I (Springer Lecture Note 192) p.207-253.

[M1] Harvard notes on topological stability (1970).

[M2] 'Stratifications and Mappings' in Dynamical Systems, M.M. Peixoto (ed.) p.195-223, Academic Press (1973).

[T1] R. Thom, La stabilité topologique des applications polynomiales, L'Enseignement Math. 8 p.24-33 (1962).

[T2] 'Local topological properties of differentiable mappings' in Differential Analysis, p.191-202, Oxford U.P. (1964).

[T3] Ensembles et Morphismes Stratifiés, Bull. Amer. Math. Soc. 75 , p.240-284 (1969).

[T4] 'The bifurcation subset of a space of maps' in Manifolds-Amsterdam 1970 (Springer Lecture Note 197) p.202-208.

[TL] R. Thom and H. Levine, 'Singularities of differential mappings' reprinted in Proc. Liverpool Symp. I (Springer Lecture Note 192) p.1-89.

[V] A.N. Varchenko, Algebro-Geometrical Equisingularity and Local Topological Classification of Smooth Mappings, Proc. of the Int. Congress of Math. Vancouver, 1974, p.427-431.

[W1] H. Whitney, On singularities of mappings of Euclidean spaces I, Mappings of the plane into the plane, Ann. of Math. 62 , p.374-410 (1955).

[W2] Tangents to an analytic variety, Ann. of Math. 81 p.469-549 (1965).

CHAPTER I

Construction of Canonical Stratifications

Christopher Gibson

§1 Whitney Stratified Sets

For the purposes of this volume a <u>stratification</u> of a subset V of a smooth manifold M is a partition $\mathcal{S}$ of V into smooth submanifolds of M (called the <u>strata</u>) which satisfies the

<u>Local Finiteness Condition</u>

Every point in V has a neighbourhood in M which meets only finitely many strata.

A good example to keep in mind is that of an algebraic set $V \subseteq \mathbb{R}^m$ (i.e. a set defined by the vanishing of finitely many polynomials). Recall that the set ΣV of all singular points of V is another algebraic set (of strictly lower dimension) and that $V - \Sigma V$ is a smooth manifold. One obtains a filtration $V_d \supseteq V_{d-1} \supseteq \ldots\ldots$ of V by taking $V_d = V$, where $d = \dim V$, and defining V_{i-1} to be ΣV_i if $\dim V_i = i$, and to be V_i if $\dim V_i < i$. There are finitely many differences $V_i - V_{i-1}$, each a smooth manifold of dimension i (or empty) yielding a finite stratification of V.

A simple illustration of this naive construction is provided by the algebraic set $V \subseteq \mathbb{R}^3$ defined by $x^2 = zy^2$ (the <u>Whitney umbrella</u>). The resulting stratification (into a surface with two connected components, and a line) is indicated in Fig. 1.

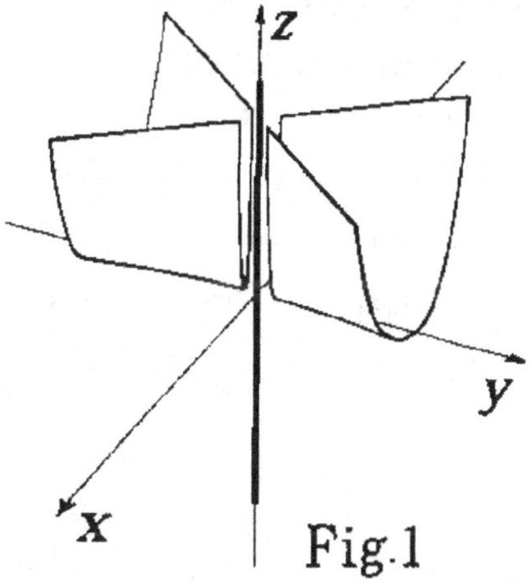

Fig.1

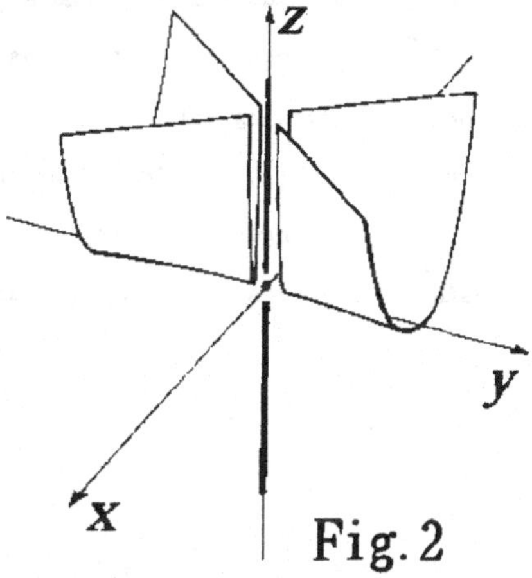

Fig.2

Ideally (returning to the general situation) one seeks stratifications for which the local topological type of the pair (M, V) is constant on each stratum. From this point of view the stratification of Fig. 1 is unsatisfactory. Indeed, if we take a point on the z-axis and sketch the intersection of a small ball centred at that point with V we obtain

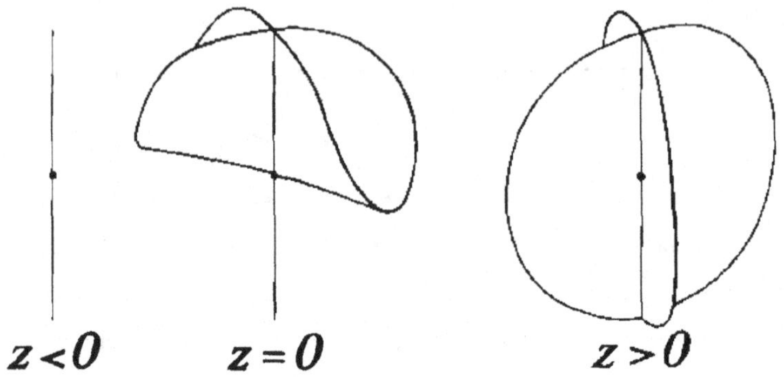

Clearly, something rather special happens at the origin, where the local topological type changes. If we agree to split up the z-axis into $z > 0$, $z = 0$, $z < 0$ we obtain a second stratification of the Whitney umbrella (indicated in Fig. 2) in which the drawback of our first stratification is no longer present. The question of just what happens at a point on a stratum where the local topological type changes was analyzed by [Whitney 1]. To avoid such situations as that in Fig. 1 he introduced the following condition for smooth submanifolds X^p, Y^q of a smooth manifold M^m, and a point $x \in X$. Take first the case when $M \subseteq \mathbb{R}^m$ is an open set. One says that <u>Y is Whitney regular over X at x</u> when the following holds. If (x_i), (y_i) are sequences in X, Y respectively (with $x_i \neq y_i$), both converging to x, if the sequence of tangent spaces $(T_{y_i}Y)$ converges to a subspace $T \subseteq \mathbb{R}^m$ (in the Grassmannian of q-dimensional subspaces of $\mathbb{R}^m$), and if the sequence $(\overrightarrow{x_i y_i})$ of lines containing

$x_i - y_i$ converges to a line $L \subseteq \mathbb{R}^m$ (in the Grassmannian of 1-dimensional subspaces of $\mathbb{R}^m$), then $L \subseteq T$. (On several occasions we shall tacitly use the fact that then automatically we have $T_x X \subseteq T$.) One checks easily the following. Given a diffeomorphism of $M \subseteq \mathbb{R}^m$ onto another open set $M' \subseteq \mathbb{R}^m$ mapping X, Y, x to X', Y', x' respectively, Y is Whitney regular over X at x if and only if Y' is Whitney regular over X' at x'. It follows that we can define Whitney regularity of Y over X at x in the case when M is an arbitrary smooth manifold. And we say that _Y is Whitney regular over X_ when it is so at every point in X.

A _Whitney stratification_ of a subset V of a smooth manifold M is a stratification $\mathcal{S}$ of V which satisfies the

Whitney Regularity Condition

Any stratum $Y \in \mathcal{S}$ is Whitney regular over any other stratum $X \in \mathcal{S}$.

The reader will easily convince himself that the stratification of the Whitney umbrella in Fig. 1 is not a Whitney stratification (the surface fails to be Whitney regular over the z-axis at the origin). On the other hand the stratification of Fig. 2 is a Whitney stratification. In the course of Chapter II it will be verified that Whitney regularity is sufficient for our purposes; indeed that on a connected component of a stratum of a Whitney stratification the local topological type remains constant.

For the purposes of this volume it will be convenient to agree that a _Whitney stratified set_ is a pair $(V, \mathcal{S})$ with V a subset of a smooth manifold M, and $\mathcal{S}$ a Whitney stratification of V.

Before proceeding any further we shall establish the few facts we shall require concerning Whitney stratifications. The first is the following lemma of [Mather 7].

(1.1) Let X, Y be smooth submanifolds of a smooth manifold M, let $x \in X \cap \overline{(Y - X)}$, and let Y be Whitney regular over X at x : then $\dim X < \dim Y$.

Proof We can suppose $M = \mathbb{R}^m$. Choose a sequence (y_i) in $Y - X$ converging to x. Compactness of the Grassmannian allows us to assume that $(T_{y_i} Y)$ converges to a subspace $T \subseteq \mathbb{R}^m$ (of the same dimension as Y). For sufficiently large i there is a point $x_i \in X$ which minimizes the distance to y_i. And, again using compactness of the Grassmannian, we can suppose that the lines $x_i y_i$ converge to a line L. Since Y is Whitney regular over X at x we have $L \subseteq T$. Also, we shall have $T_x X \subseteq T$, as was observed above, so that $T_x X + L \subseteq T$. But the line $x_i y_i$ is perpendicular to $T_{x_i} X$, so L is perpendicular to $T_x X$. Thus $\dim X = \dim T_x X < \dim T = \dim Y$.

Q. E. D.

The next three results are all concerned with the subject of gaining new Whitney stratifications from old. First of all we can take products. Let $\mathcal{X}_1, \ldots, \mathcal{X}_n$ be stratifications of subsets $V_1, \ldots, V_n$ of smooth manifolds. Clearly, we obtain a stratification of $V_1 \times \ldots \times V_n$ by taking its strata to be sets of the form $X_1 \times \ldots \times X_n$ with $X_i \in \mathcal{X}_i$ for $1 \leq i \leq n$: we call this the product stratification, and denote it $\mathcal{X}_1 \times \ldots \times \mathcal{X}_n$.

(1.2) Let $\mathcal{X}_1, \ldots, \mathcal{X}_n$ be Whitney stratifications of subsets $V_1, \ldots, V_n$ of smooth manifolds; the product stratification $\mathcal{X}_1 \times \ldots \times \mathcal{X}_n$ is a Whitney stratification $V_1 \times \ldots \times V_n$.

The proof is sufficiently obvious to justify its omission (using no more than compactness of the Grassmannian). Another way of gaining new stratifications is as follows. Recall first that smooth submanifolds $X_1, \ldots, X_n$ of a smooth manifold M are in general position when the natural map $T_x M \to \oplus T_x M / T_x X_i$ is surjective for each point x in the intersection. Now let $\mathcal{X}_1, \ldots, \mathcal{X}_n$ be stratifications of subsets $V_1, \ldots, V_n$ of a smooth manifold M.

We say that $\mathcal{X}_1, \ldots, \mathcal{X}_n$ are in <u>general position</u> when $X_1, \ldots, X_n$ are in general position for all $X_1 \in \mathcal{X}_1, \ldots, X_n \in \mathcal{X}_n$. And in that case we denote by $\mathcal{X}_1 \cap \ldots \cap \mathcal{X}_n$ the stratification of $V_1 \cap \ldots \cap V_n$ obtained by taking the strata to be sets $X_1 \cap \ldots \cap X_n$ with $X_1 \in \mathcal{X}_1, \ldots, X_n \in \mathcal{X}_n$.

(1.3) <u>Let $\mathcal{X}_1, \ldots, \mathcal{X}_n$ be Whitney stratifications of subsets</u> $V_1, \ldots, V_n$ <u>of a smooth manifold</u> M. <u>If</u> $\mathcal{X}_1, \ldots, \mathcal{X}_n$ <u>are in general position then</u> $\mathcal{X}_1 \cap \ldots \cap \mathcal{X}_n$ <u>is a Whitney stratification of</u> $V_1 \cap \ldots \cap V_n$.

<u>Proof</u> We can suppose $M = \mathbb{R}^m$. Let $X = X_1 \cap \ldots \cap X_n$ and $Y = Y_1 \cap \ldots \cap Y_n$ be strata in $\mathcal{X}_1 \cap \ldots \cap \mathcal{X}_n$ with $X_s, Y_s \in \mathcal{X}_s$ for $1 \leq s \leq n$. Let $x \in X$, and let (x_i), (y_i) be sequences in X, Y respectively converging to x for which the lines $\overrightarrow{x_i y_i}$ converge to a line L, and $(T_{y_i} Y)$ converges to a subspace $T \subseteq \mathbb{R}^m$ (of the same dimension as Y). We have to show $L \subseteq T$. Using compactness of the Grassmannian we can suppose that the sequence $(T_{y_i} Y_s)$ converges to a subspace T_s (of the same dimension as Y_s) for $1 \leq s \leq n$. Since each Y_s is Whitney regular over X_s at x we have $L \subseteq T_s$ for $1 \leq s \leq n$, and hence $L \subseteq \bigcap T_s$. We will show that $T = \bigcap T_s$, which will clinch the result. By hypothesis the subspaces $T_{y_i} Y_1, \ldots, T_{y_i} Y_n$ are in general position (with intersection $T_{y_i} Y$) for all i: it follows that $T \subseteq \bigcap T_s$, using an obvious lemma concerning convergence in the Grassmannian. By previous remarks we automatically have $T_s \supseteq T_x X_s$ for $1 \leq s \leq n$. But by hypothesis the subspaces $T_x X_1, \ldots, T_x X_n$ are in general position, and hence $T_1, \ldots, T_n$ are as well. That implies $\dim T = \dim \bigcap T_s$, and it follows that $T = \bigcap T_s$.

Q. E. D.

One final method of producing new stratifications is as follows. Let $\mathcal{X}'$ be a stratification of a subset V' of a smooth manifold M', and let $f : M \to M'$ be a smooth mapping <u>transverse</u> to $\mathcal{X}'$ (i.e. transverse to all the strata of $\mathcal{X}'$). Clearly, we obtain a stratification $\mathcal{X}$ of $f^{-1}V'$ by taking the strata to be sets $f^{-1}X'$ with $X' \in \mathcal{X}'$: we call $\mathcal{X}$ the <u>induced stratification</u> on $f^{-1}X'$.

(1.4) Let $f : M \to M'$ be a smooth mapping transverse to a stratification $\mathcal{X}'$ of a subset V' of M', and let $\mathcal{X}$ be the induced stratification on $f^{-1}V'$: if $\mathcal{X}'$ is a Whitney stratification then so too is $\mathcal{X}$.

Proof   Factorize f as the composite $M \xrightarrow{F} \text{graph } f \xrightarrow{\pi} M'$ with $F(x) = (x, f(x))$ and π the restriction to graph f of the projection $M \times M' \to M'$. F maps $\mathcal{X}$ diffeomorphically to a stratification $\mathcal{X}''$ of a subset of graph f, and it suffices to show that $\mathcal{X}''$ is a Whitney stratification. To this end observe that by (1.2) the product stratification on $M \times V'$ is a Whitney stratification, which by hypothesis is transverse to graph f. It follows from (1.3) that the naturally induced stratification on $(M \times V') \cap$ graph f is a Whitney stratification. But this latter stratification is $\mathcal{X}''$, and we are done.

Q. E. D.

A particular case of the above is provided by the following situation. Let $\mathcal{X}$ be a stratification of a subset V of a smooth manifold M, and let $U \subseteq M$ be open. The inclusion $U \to M$ is automatically transverse to $\mathcal{X}$, so there is an induced stratification on $U \cap V$ which we denote by $\mathcal{X}|U$ and call the restriction of $\mathcal{X}$ to U. And by (1.4) if $\mathcal{X}$ is a Whitney stratification then the restriction $\mathcal{X}|U$ is likewise a Whitney stratification.

The usefulness of Whitney stratifications depends on being able to find a sufficiently large class of sets for which they exist. In fact it is not merely existence which we seek; we require also uniqueness in the sense of a definition to be given below.

First however a word of explanation may be welcome, since the definition we intend to give is not the obvious one. Consider all Whitney stratifications of a given subset V of a smooth manifold M. These are partially ordered by refinement (if $\mathcal{X}, \mathcal{X}'$ are stratifications of V then $\mathcal{X}$ refines $\mathcal{X}'$ when every stratum of $\mathcal{X}$ is contained in a stratum of $\mathcal{X}'$). One might hope to obtain a greatest element (i.e. one which is refined by any other) under this partial order, but this is crying for the moon, as the following example shows. Let $V \subseteq \mathbb{R}^2$ be the set comprising the origin and the set defined by $y^2 < x^2$.

These yield a Whitney stratification $\mathcal{I}$ of V, and it is easy to see that if the Whitney stratifications of V have a greatest element under refinement then $\mathcal{I}$ must be it. But the x-axis and its complement in V also yield a Whitney stratification $\mathcal{I}'$ of V which does not refine $\mathcal{I}$. We conclude that there is no greatest element.

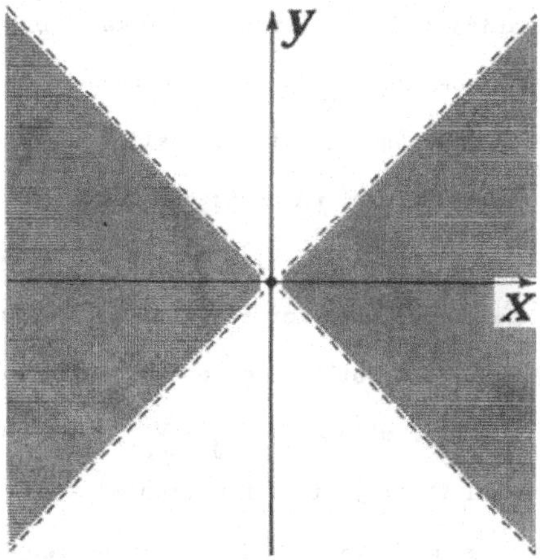

With the above in mind we follow [Mather 8] in introducing the next definition. Given a subset V of a smooth manifold M, and a stratification $\mathcal{I}$ of V we define the <u>associated filtration of V by dimension</u> to be the filtration (V_i) of V obtained by taking V_i to be the union of the strata in $\mathcal{I}$ of dimension $\leq i$. We can now introduce a rather different partial order on the stratifications of V. Let $\mathcal{I}$, $\mathcal{I}'$ be two such, and let (V_i), (V_i') be the associated filtrations by dimension. We write $\mathcal{I} < \mathcal{I}'$ when there exists an integer i for which $V_i \subset V_i'$ but $V_j = V_j'$ for all $j > i$. In this way we induce a partial order $\leq$ on the Whitney stratifications of V. If this partial order has a least element we call it a <u>minimal Whitney stratification</u> of V.

Notice incidentally the following application of (1.1). Let $\mathcal{I}$ be a Whitney stratification of a subset V of a smooth manifold M, and let (V_i) be the associated filtration by dimension; then

(i) $V_i - V_{i-1}$ is empty, or a smooth submanifold of M of dimension i, and

(ii) $V_i - V_{i-1}$ does not meet the closure of $V_j - V_{j-1}$ unless $i \leq j$.

For the sets which arise naturally in this area of mathematics there is little doubt that the "minimal Whitney stratification" is the correct notion to aim for. But for arbitrary sets the notion has a drawback. Suppose that V is a subset of a smooth manifold M admitting a minimal Whitney stratification $\mathcal{S}$, and that $U \subseteq M$ is an open set : there is no reason to suppose that the restriction $\mathcal{S}|U$ is the minimal Whitney stratification of $V \cap U$. We shall therefore introduce a stronger notion, which is obviously local. Let $\mathcal{S}$ be a Whitney stratification of V , and let (V_i) be the associated filtration by dimension. We call $\mathcal{S}$ <u>canonical</u> when for each index i the set $V_i - V_{i-1}$ is the largest subset of V_i which firstly is a smooth submanifold of M of dimension i , and secondly over which the $V_j - V_{j-1}$ are Whitney regular for $j > i$. Clearly a canonical Whitney stratification of V is necessarily minimal. And for the sets we have in mind it will follow from the very construction of the canonical Whitney stratification that the two notions coincide. Notice incidentally that these are notions of differential topology in the sense that a diffeomorphism of M onto a smooth manifold M' , mapping V onto V' , will map the minimal (respectively canonical) Whitney stratification of V, if such exists, onto the minimal (respectively canonical) Whitney stratification of V' .

Before proceeding to these constructions we should say something about the definition of the term "Whitney stratification" we have adopted, since it is not the accepted one. It is usual, pursuing the analogy with cellular subdivision in topology, to insist that Whitney stratifications should in addition satisfy the

<u>Frontier Condition</u>

Let X, Y be strata with $X \cap \bar{Y} \neq 0$: then $X \subseteq \bar{Y}$ (i.e. the frontier of a stratum is a union of strata).

The Frontier Condition plays no role in the proofs of the results stated in this volume. Moreover in practice it is something of an embarrassment, since it is not preserved under natural operations on stratifications (e.g. taking intersections of stratifications in general position). For these reasons we feel it better to dispose with the Frontier Condition altogether. In this connexion it is perhaps of interest to mention the following fact, which will be proved in Chapter II. Suppose that $\mathcal{X}$ is a Whitney stratification of a locally closed subset V of a smooth manifold M. Let $\mathcal{X}^c$ be the partition of V into the connected components of the strata of $\mathcal{X}$. It turns out that $\mathcal{X}^c$ satisfies the Local Finiteness Condition, hence is a stratification of V, indeed a Whitney stratification. Moreover $\mathcal{X}^c$ satisfies the Frontier Condition. Clearly, $\mathcal{X}$ is minimal (canonical) if and only if $\mathcal{X}^c$ is.

§2 Semialgebraic Sets

We seek a useful class of sets admitting canonical Whitney stratifications. One hopes for a class large enough to include algebraic sets, and (for practical reasons) closed under as many set theoretic and topological operations as is possible. For the purposes of this volume semialgebraic sets (defined below) will suffice - though larger classes with more or less identical properties exist. As was pointed out in the introduction to this volume we shall merely quote the basic results about semialgebraic sets here, referring the interested reader to the publications of [Lojasiewicz] for proofs.

The class of semialgebraic subsets of $\mathbb{R}^m$ is defined to be the smallest Boolean algebra of subsets of $\mathbb{R}^m$ which contains all sets of the form

$$\{x \in \mathbb{R}^m : f(x) > 0\}$$

with $f : \mathbb{R}^m \to \mathbb{R}$ a polynomial function.

Let us first dispose of the trivial facts. It follows immediately from the definitions that the semialgebraic subsets of $\mathbb{R}^m$ are closed under finite unions, finite intersections, complements and products. Moreover the inverse image of a semialgebraic subset of $\mathbb{R}^p$ under a polynomial mapping $f : \mathbb{R}^n \to \mathbb{R}^p$ is semialgebraic. The first substantial fact is the Tarski -

Seidenberg Theorem.

(2.1) The image of a semialgebraic subset of $\mathbb{R}^n$ under a polynomial mapping $f : \mathbb{R}^n \to \mathbb{R}^p$ is semialgebraic.

[Lojasiewicz's] proof of (2.1) yields another basic result, generalizing the Finiteness Theorem of [Whitney 2].

(2.2) A semialgebraic set has only finitely many connected components, each of which is semialgebraic.

A delightful little application of the Tarski - Seidenberg Theorem is provided by

(2.3) The closure (hence also the interior and the frontier) of a semialgebraic set $A \subseteq \mathbb{R}^m$ is semialgebraic.

Proof Clearly $B \subseteq \mathbb{R}^m \times \mathbb{R}^m \times \mathbb{R}$ comprising all triples (x, y, ϵ) with $x \in A$ and $|x - y| < \epsilon$ is semialgebraic. Now consider the polynomial projections $\mathbb{R}^m \times (\mathbb{R}^m \times \mathbb{R}) \xrightarrow{p} \mathbb{R}^m \times \mathbb{R} \xrightarrow{q} \mathbb{R}^m$. A minor computation verifies that $\bar{A} = \mathbb{R}^m - q(\mathbb{R}^m \times \mathbb{R} - p(B))$, which is semialgebraic by (2.1).

Q. E. D.

We can introduce a notion of "dimension" for semialgebraic sets as follows. Let V be a subset of a smooth manifold M. A point $x \in V$ is regular (of dimension d) when x has a neighbourhood U in M for which $U \cap V$ is a smooth submanifold of M (of dimension d). The reader is warned that this is not quite the definition used in [Lojasiewicz] : however it can be proved that the two definitions are equivalent. Assume that V has at least one regular point; then the dimension $\dim V$ of V is defined to be the maximal dimension of a regular point. (Of course one has to adopt some convention concerning the dimension of the empty set, e.g. that it is -1.) That the dimension of a semialgebraic set is always defined follows from

(2.4) <u>Every non-void semialgebraic set has at least one regular point; in fact the regular points lie dense.</u>

It is not hard to show that for semialgebraic subsets A, B of $\mathbb{R}^m$ one has

(i) If $A \subseteq B$ then $\dim A \leqslant \dim B$.

(ii) $\dim (A \cup B) = \max (\dim A, \dim B)$.

(iii) $\dim \bar{A} = \dim A$.

The points in a semialgebraic set V which fail to be regular, of maximal dimension, are called <u>singular</u>. (Thus a point in V can be both regular and singular.) And the singular set ΣV of V is the set of all singular points of V . Notice that ΣV is closed in V . The next result lies rather deeper than the preceeding ones.

(2.5) <u>The singular set</u> ΣV <u>of a semialgebraic set</u> V <u>is semialgebraic, of dimension</u> $< \dim V$.

Now we come closer to the matter in hand. Let X, Y be smooth submanifolds of a smooth manifold M . We define B(X, Y) - the <u>bad set</u> - to be the set of points $x \in X$ where Y fails to be Whitney regular over X at x . The basic fact about the bad set is Whitney's Theorem, which lies at least as deep as any of the preceeding propositions.

(2.6) <u>Let</u> X, Y <u>be semialgebraic smooth submanifolds of</u> $\mathbb{R}^m$: <u>then</u> B(X, Y) <u>is semialgebraic, of dimension</u> $< \dim X$.

A corresponding result in the complex case appeared in a long and difficult paper of [Whitney 1] . Shortly afterwards [Thom 1] published a more illuminating account. A recent account along the same basic lines is provided by [Wall] . The reader versed in algebraic geometry is referred to the elegant proof given by [Hironaka] using his resolution of singularities.

In order to reduce the symbolism it will be convenient at this point to introduce, for any semialgebraic sets A, B in $\mathbb{R}^m$ the set

$$W(A, B) = \Sigma A \cup B(A - \Sigma A, B - \Sigma B) \; .$$

It follows from the preceeding results that $W(A, B)$ is semialgebraic, of dimension $< \dim A$.

(2.7) <u>Any semialgebraic set</u> $V \subseteq \mathbb{R}^m$ <u>admits a canonical Whitney</u> <u>stratification</u> $\mathfrak{X}$ <u>having finitely many semialgebraic strata.</u>

<u>Proof</u>   The stratification of an algebraic set described at the beginning of §1 works equally well for a semialgebraic set.   What we do is to modify the construction by deleting from any proposed stratum the (closure of the) set of points where previously defined strata fail to be Whitney regular over it. Formally, we proceed as follows.   Let $d = \dim V$.   We construct a filtration $V = V_d \supseteq V_{d-1} \supseteq \dots$ of V by semialgebraic sets V_i closed in V with $\dim V_i \leqslant i$ and each difference $V_i - V_{i-1}$ a smooth manifold of dimension i (or empty).   Suppose inductively that $V_d, V_{d-1}, \dots, V_i$ have been constructed according to this prescription.   If $\dim V_i < i$ we put $V_{i-1} = V_i$ and are done. If $\dim V_i = i$ we put

$$V_{i-1} = \begin{array}{c} \text{closure} \\ \text{in } V \text{ of} \end{array} \left\{ \bigcup_{j=i+1}^{d} W(V_i, V_j - V_{j-1}) \right\} .$$

That V_{i-1} is semialgebraic, of dimension $\leqslant (i - 1)$, follows from the preceeding remarks.   And $V_i - V_{i-1}$ is a smooth manifold since it is obtained from the smooth manifold $V_i - \Sigma V_i$ by deleting a closed set.   We take $\mathfrak{X}$ to be the stratification of V whose strata are the $V_i - V_{i-1}$.   It follows from the construction that $\mathfrak{X}$ is a Whitney stratification.   And of course it has finitely many semialgebraic strata.   It remains to show that $\mathfrak{X}$ is canonical, which fact again follows immediately from the construction.

Q. E. D.

Since the literature on Whitney stratified sets is rather devoid of examples this may be an appropriate point to digress from the main line of thought and describe a few examples culled from the folk-lore of the subject.   In the few known situations where the Whitney Regularity Condition has been verified the

Whitney Theorem on the bad set has been the main tool. A particularly nice set-up is provided by a semialgebraic stratification (i.e. one with semialgebraic strata) of a set $V \subseteq \mathbb{R}^m$ which enjoys the following homogeneity property; that given any two points on the same stratum there exists a diffeomorphism of a neighbourhood in $\mathbb{R}^m$ of the one point onto a neighbourhood in $\mathbb{R}^m$ of the other point which preserves strata. It follows immediately from the Whitney Theorem that such a stratification must be a Whitney stratification.

Take for instance a Lie group action on a smooth manifold $V \subseteq \mathbb{R}^m$ having finitely many semialgebraic orbits. Here the orbits enjoy an obvious homogeneity property as a result of which they are non-singular semialgebraic sets, using (2.4), hence smooth submanifolds of V providing a stratification. And the same homogeneity property ensures that this is a Whitney stratification. By way of explicit illustration, let V, W be finite - dimensional real vector spaces, and let Σ^i be the subset of $\mathrm{Hom}(V, W)$ comprising all linear maps $V \to W$ having kernel rank i : the Σ^i provide a Whitney stratification of $\mathrm{Hom}(V, W)$ since they are the orbits under the natural action of the Lie group $GL(V) \times GL(W)$, and clearly semialgebraic.

A similar situation yielding examples of Whitney stratifications is provided by a smooth action of a compact Lie group G on a smooth manifold M. Recall that locally there are only finitely many orbit types, and that the union of all orbits having a given type is a smooth submanifold of M. Thus these unions provide a stratification of M. And using slices it is not difficult to establish the Whitney Regularity Condition.

Here are two further examples where positive work is required to verify the statements. Consider the partition of $\mathbb{R}^{n+1}$ obtained by separating points $a = (a_0, a_1, \ldots, a_n)$ according to the number and the multiplicities of the complex zeros of the polynomial $a_0 z^n + a_1 z^{n-1} + \ldots + a_n$: this provides a finite semialgebraic Whitney stratification. Finally, an example drawn from linear algebra : the partition of the set of endomorphisms of a finite dimensional real vector space by Segre symbol provides another finite semialgebraic Whitney stratification. A proof can be found in [Gibson].

§3 Thom Stratified Mappings

Let $f : N \to P$ be a smooth mapping, and let $A \subseteq N$, $B \subseteq P$ be sets with $f(A) \subseteq B$. A <u>stratification</u> of $f : A \to B$ is a pair $(\mathcal{X}, \mathcal{X}')$ with $\mathcal{X}$, $\mathcal{X}'$ Whitney stratifications of A, B respectively for which the following conditions are satisfied.

(i) f maps strata into strata (but not necessarily onto).

(ii) Let X be a stratum of $\mathcal{X}$, mapped by f into a stratum X' of $\mathcal{X}'$: then $f : X \to X'$ is a submersion.

A simple example of a stratified mapping is provided by the Whitney cusp mapping of the plane, i.e. the mapping $f : \mathbb{R}^2 \to \mathbb{R}^2$ given by $(x, y) \to (u, v)$ with $u = x$, $v = y^3 - xy$. The set of critical points is the parabola $x = 3y^2$, with image under f the cuspidal cubic $4u^3 - 27v^2 = 0$. And the inverse image of this curve is the union of the parabola with another - see below.

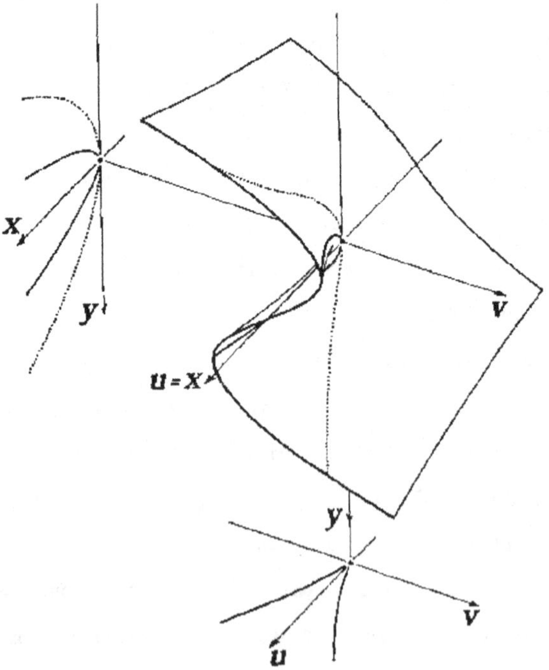

There are certainly circumstances where the mere existence of a stratification for a smooth mapping provides useful topological information. However, for the purposes of the Topological Stability Theorem we require a little more - precisely, we require stratifications of smooth mappings to which Thom's Second Isotopy Lemma applies. To this end we introduce a condition on stratifications of smooth mappings analogous to the Whitney Regularity Condition.

Let $f : N \to P$ be a smooth mapping, let X, Y be smooth submanifolds of N for which the restrictions $f|X$ and $f|Y$ have constant rank, and let $x \in X$. We say that *Y is Thom regular over X at x relative to f* when the following holds. Let (y_i) be a sequence of points in Y converging to x such that $(\ker T_{y_i}(f|Y))$ converges, in the appropriate Grassmann bundle, to a plane T; then $\ker T_x(f|X) \subseteq T$. And we say that *$Y$ is Thom regular over X relative to f* when it is so at every point $x \in X$.

Now let $f : N \to P$ be a smooth mapping, let $A \subseteq N$, $B \subseteq P$ be sets with $f(A) \subseteq B$, and let $(\mathcal{S}, \mathcal{S}')$ be a stratification of $f : A \to B$. We call $(\mathcal{S}, \mathcal{S}')$ a *Thom stratification* when it satisfies the

Thom Regularity Condition

For any strata $X, Y \in \mathcal{S}$ we have Y Thom regular over X relative to f.

For the purposes of this volume it will be convenient to agree that a Thom stratified mapping is a triple $(f, \mathcal{S}, \mathcal{S}')$ with f a smooth mapping, and $(\mathcal{S}, \mathcal{S}')$ a Thom stratification for f. It is customary to reseve the term *Thom mapping* for a smooth mapping which admits a Thom stratification.

An example where the Thom Regularity Condition has a simple geometric meaning is provided by the smooth function $f : \mathbb{R}^2 \to \mathbb{R}$ given by $f(x, y) = xy$. f admits an obvious stratification, indicated in the figure below. Suppose we take a sequence of points (v_i) in the open first quadrant converging to a point u on the positive x-axis. The Thom Regularity Condition merely asserts that the tangent lines at the v_i to the fibres through v_i must converge to the x-axis. Clearly then we have a Thom stratification of f.

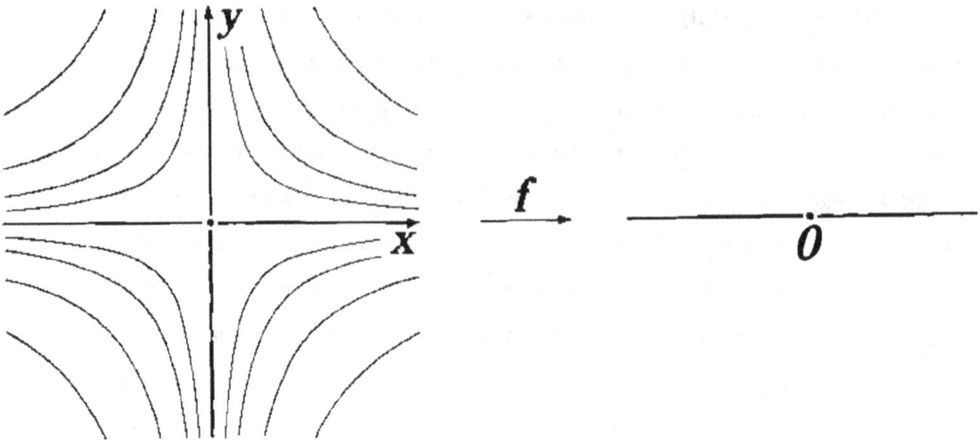

In this last example we have a natural stratification for the mapping which automatically satisfies the Thom Regularity Condition. That this is not always the case is illustrated by the pinch map $f : \mathbb{R}^2 \to \mathbb{R}^2$ given by $(x, y) \to (u, v)$ where $u = x$, $v = xy$. (It "pinches" the plane in such a way that the y-axis collapses to a point : algebraic geometers will recognise it as the σ-process). Here again there is a natural stratification of f, indicated in the figure below.

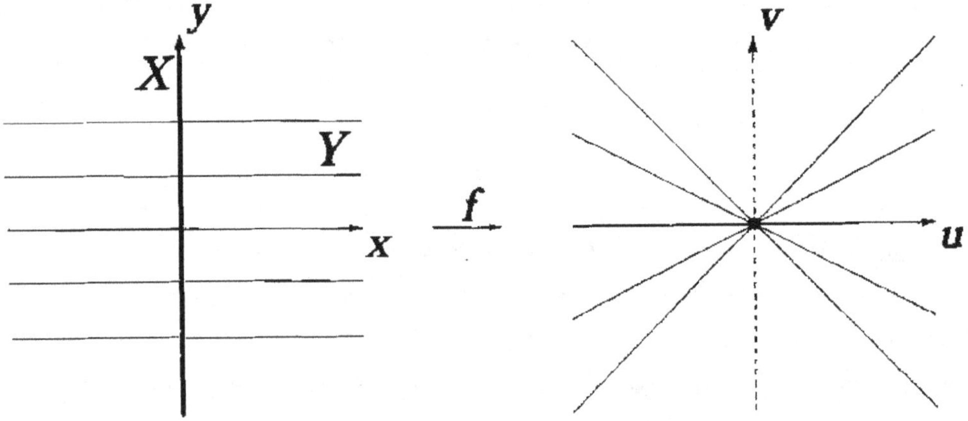

The strata in the domain are X = critical set of f = y-axis, and its complement Y. Take $x \in X$, and a sequence (y_i) in Y converging to x. Clearly $\ker T_{y_i}(f|Y) = \{0\}$, so the sequence of kernels converges to $\{0\}$. Were the Thom Regularity Condition to hold we should have $\ker T_x(f|X) \subseteq \{0\}$, hence $= \{0\}$. But this is false since the kernel is X .

In fact we can improve upon this argument to show that <u>the pinch map f does not admit a Thom stratification</u>. For suppose it did admit one $(\mathcal{F}, \mathcal{F}')$. I claim that for any stratum $U \in \mathcal{F}$ the intersection $U \cap X$ is countable. Certainly $\dim U \leq 2$. The case $\dim U = 0$ is trivial; and the case $\dim U = 2$ does not arise since otherwise if $U \cap X \neq \emptyset$ the pinch map f could not map U submersively into a stratum in $\mathcal{F}'$. We are left with the case $\dim U = 1$. It suffices to show that U, X intersect transversally (since then the intersection is a smooth submanifold of $\mathbb{R}^2$ of dimension 0, hence countable) . So choose $x \in U \cap X$. We can certainly find a sequence (y_i) in $\mathbb{R}^2 - X - U$ converging to x . Since $\mathcal{F}$ is locally finite there will be a stratum V containing a subsequence - which we also denote (y_i) - converging to x . All the $\ker T_{y_i}(f|V)$ are trivial, so the Thom Regularity Condition implies that $\ker T_x(f|U)$ is trivial. And that entails that U, X intersect transversally at x . Finally, observe that we must have $X = \bigcup_{U \in \mathcal{F}} U \cap X$, which provides the required contradiction since the left-hand side is uncountable, but the right-hand side is countable (being the countable union of countable sets) .

For the purposes of the Topological Stability Theorem one wishes to associate Thom stratifications with infinitesimally stable mappings, and since these locally look like polynomial mappings (a statement which will be made precise in Chapter III) the natural starting point is to try and associate Thom stratifications with polynomial mappings. The example of the pinch map makes it clear that we must impose some kind of restriction, preferably one which is "generic" . We therefore impose on our mappings the

<u>Genericity Condition</u>

The restriction $f|\Sigma$ of $f : N \to P$ to its critical set Σ is both proper and finite-to-one.

In Chapter III it will be shown that this condition is indeed satisfied by some representative of any infinitesimally stable map germ, so is "generic".

We come now to the explicit construction of Thom stratifications for generic polynomial mappings. Our constructions will be simplified by the following idea of Looijenga. Let $f : N \to P$ be a smooth mapping. We write $\Sigma = \Sigma(f)$ for the set of critical points of f, and $C = C(f)$ for the set of critical values. Notice that if f satisfies the Genericity Condition then $C(f)$ is closed. A <u>partial stratification</u> for f is a Whitney stratification $\mathcal{C}$ of C which satisfies the following conditions for any strata U, V of $\mathcal{C}$ (possibly equal).

(PS1) $f^{-1}U \cap \Sigma$ is a smooth submanifold of N, and the restriction $f : f^{-1}U \cap \Sigma \to U$ is a local diffeomorphism.

(PS2) $f^{-1}V \cap \Sigma$ is Whitney regular over $f^{-1}U \cap \Sigma$.

(PS3) $f^{-1}V - \Sigma$ is Whitney regular over $f^{-1}U \cap \Sigma$.

Now recall from §1 that there is a partial order $\leq$ on the stratifications of C, inducing a partial order on the partial stratifications of f. If the induced partial order has a least element we call it a <u>minimal partial stratification</u> of f. As in the simpler situation discussed in §1 there is no reason to suppose that this is a local notion. We introduce therefore a stronger notion. Let $\mathcal{C}$ be a partial stratification for f, and let (C_i) be the associated filtration by dimension. Certainly then the $M_i = C_i - C_{i-1}$ satisfy the following conditions, for each j.

(PS0) M_j is empty, or a smooth submanifold of P of dimension j over which each M_k is Whitney regular for $k > j$.

(PS1) $f^{-1}M_j \cap \Sigma$ is a smooth submanifold of N, and the restriction $f : f^{-1}M_j \cap \Sigma \to M_j$ is a local diffeomorphism.

(PS2) $f^{-1}M_k \cap \Sigma$ is Whitney regular over $f^{-1}M_j \cap \Sigma$ for $k > j$.

(PS3) $f^{-1}M_k - \Sigma$ is Whitney regular over $f^{-1}M_j \cap \Sigma$ for $k \geq j$.

We shall call $\mathcal{C}$ a <u>canonical partial stratification</u> of f when M_j is the largest subset of C_j for which the above conditions hold, for each index j. Clearly a canonical partial stratification of f is necessarily minimal. And for the mappings we have in mind it will follow from the very construction of canonical partial stratifications that the two notions coincide. Without more ado we shall come to the point of partial stratifications.

(3.1) Let $f : N \to P$ be a smooth mapping whose set C of critical values is closed : if f admits a partial stratification $\mathscr{C}$ then f admits a Thom stratification.

Proof Fist, note that we obtain a Whitney stratification $\mathscr{S}'$ of P by augmenting $\mathscr{C}$ by $(P - C)$, which is open in P, hence a smooth submanifold of P : since $(P - C)$ is open in P it will automatically be Whitney regular over any stratum of $\mathscr{C}$. And clearly we obtain a stratification $\mathscr{S}$ of N by taking its strata to be sets of the form $f^{-1}(P - C)$, $f^{-1}W \cap \Sigma$, $f^{-1}W - \Sigma$ with W a stratum of $\mathscr{S}'$, and Σ the critical set of f. I claim first that $\mathscr{S}$ is a Whitney stratification of N. To this end let X, Y be strata in $\mathscr{S}$.

From the very construction of $\mathscr{S}$ it follows that there are strata U, V in $\mathscr{S}'$ with $f(X) \subseteq U$, $f(Y) \subseteq V$. Since any stratum in $\mathscr{S}$ is contained in Σ, or disjoint from Σ, there are four possible cases to consider.

(i) $X \cap \Sigma = \emptyset$ and $Y \subseteq \Sigma$ Since Σ is closed we have $X \cap \bar{Y} = \emptyset$ and Whitney Regularity follows trivially.

(ii) $X \cap \Sigma = \emptyset$ and $Y \cap \Sigma = \emptyset$ $f|N - \Sigma$ is a submersion so certainly transverse to U, V. It follows from (1.4) that $f^{-1}V - \Sigma$ is Whitney regular over $f^{-1}U - \Sigma$, so Y is Whitney regular over X.

(iii) $X \subseteq \Sigma$ and $Y \subseteq \Sigma$ In this case Y is Whitney regular over X by (PS2).

(iv) $X \subseteq \Sigma$ and $Y \cap \Sigma = \emptyset$ One possibility is that $Y = f^{-1}(P - C)$ in which case Y (being open) is certainly Whitney regular over X : the remaining possibility is covered by (PS3).

Next, I claim that $(\mathscr{S}, \mathscr{S}')$ is a stratification of f. By construction f maps strata into strata. We have to show that it does so submersively. For strata in $\mathscr{S}$ of the form $f^{-1}(P - C)$ or $f^{-1}W - \Sigma$, with W a stratum of $\mathscr{S}'$, this is immediate since $f \mid N - \Sigma$ is a submersion. And for strata in $\mathscr{S}$ of the form $f^{-1}W \cap \Sigma$ this follows from (PS1).

It remains to show that the stratification $(\mathcal{I}, \mathcal{I}')$ for f satisfies the Thom Regularity Condition. Take strata X, Y in $\mathcal{I}$. Let $x \in X$, and let (y_i) be a sequence of points in Y converging to x. As before X, Y map into strata U, V of $\mathcal{I}'$. And again we consider cases.

(i) $X \subseteq \Sigma$. By (PS1) the restriction $f : X \to U$ is a local diffeomorphism, so $\ker T_x(f|X) = \{0\}$, and trivially Y is Thom regular over X at x relative to f.

(ii) $X \cap \Sigma = \emptyset$ Necessarily $Y \cap \Sigma = \emptyset$ as well. On a neighbourhood of the point x we have $X = f^{-1}f(X)$ and $Y = f^{-1}f(Y)$. And since f is submersive outside Σ we have

$$\ker T_{y_i}(f|Y) = T_{y_i}(f^{-1}f(y_i))$$

$$\ker T_x(f|X) = T_x(f^{-1}f(x))$$

so the former converge to the latter, again showing that Y is Thom regular over X at x relative to f.

Q. E. D.

The Thom stratification $(\mathcal{I}, \mathcal{I}')$ constructed in the above proof will be called the Thom stratification <u>associated</u> to the partial stratification $\mathcal{G}$. Thus the problem of constructing a Thom stratification for a generic polynomial mapping reduces to that of constructing a partial stratification. Since this is merely a special type of Whitney stratification of the set of critical values it is to be expected that the construction will be no more than a complicated version of that described in (2.7). And indeed this will be the case. In order to make the construction work we need two preliminary results.

(3.2) <u>Let</u> $f : \mathbb{R}^n \to \mathbb{R}^p$ <u>be a polynomial mapping, and let</u> $X \subseteq \mathbb{R}^n$ <u>be semialgebraic : then</u> $\dim f(X) \leq \dim X$.

<u>Proof</u> Put $Y = f(X)$, which is of course semialgebraic by the Tarski - Seidenberg Theorem. We proceed by induction on $\dim X$. The case $\dim X = 0$ is clear, since a semialgebraic set has zero dimension if and only if it is finite.

Suppose dim X is positive, and that the result holds for semialgebraic sets of dimension < dim X. By (2.5) the singular set ΣX is semialgebraic, of dimension < dim X. The Tarski – Seidenberg Theorem tells us that $f(\Sigma X)$ is semialgebraic, as is its closure, by (2.3). Using the induction hypothesis $\dim \overline{f(\Sigma X)} = \dim f(\Sigma X) \leq \dim (\Sigma X) < \dim X$.

Let us suppose $\dim Y > \dim X$. The set $Y_1 = Y - \Sigma Y - \overline{f(\Sigma X)}$ is a semialgebraic smooth submanifold of $\mathbb{R}^p$ of dimension dim Y, since it is obtained from another, $Y - \Sigma Y$, by deleting a closed semialgebraic set of lower dimension. Also, $X_1 = X - f^{-1}(\Sigma Y \cup \overline{f(\Sigma X)})$ is a semialgebraic smooth submanifold of $\mathbb{R}^n$ of dimension dim X, since it is obtained from another, $X - \Sigma X$, by deleting a proper closed subset. In this way we obtain a surjective smooth mapping $f : X_1 \to Y_1$ with $\dim Y_1 > \dim X_1$. A point in Y_1 is a regular value of $f : X_1 \to Y_1$ if and only if it does not lie in the image $f(X_1)$. But f has at least one regular value – by Sard's Theorem, for instance – contradicting surjectivity of $f : X_1 \to Y_1$. We conclude that indeed $\dim Y \leq \dim X$.

Q. E. D.

(3.3) Let $f : \mathbb{R}^n \to \mathbb{R}^p$ be a polynomial mapping, and let $X \subseteq \mathbb{R}^n$ be semialgebraic, with $f|X$ finite-to-one: then $\dim f(X) = \dim X$.

Proof Again, $Y = f(X)$ is semialgebraic. This time we proceed by induction on dim Y. When dim Y = 0 the result is clear. Suppose dim Y is positive, and that the result holds for lower dimensions. By (2.5) the singular set ΣY is semialgebraic, of dimension < dim Y. Using the induction hypothesis, and (3.2), we have

$$\dim (f|X)^{-1}(\Sigma Y) = \dim (\Sigma Y) < \dim Y \leq \dim X .$$

Now $X_1 = X - \Sigma X - (f|X)^{-1}(\Sigma Y)$ is a smooth submanifold of $\mathbb{R}^n$ of dimension dim X, since it is obtained from another, $X - \Sigma X$, by deleting a closed semialgebraic subset of lower dimension. And $Y_1 = Y - \Sigma Y$ is a smooth submanifold of $\mathbb{R}^p$ of dimension dim Y. In this way we obtain a finite-to-one smooth mapping $f : X_1 \to Y_1$. By shrinking X_1 we can suppose $f : X_1 \to Y_1$

has constant rank. Assume $\dim X > \dim Y$. It will then be possible to construct a non-trivial smooth vector field ξ on X_1 with $\xi(X_1) \subseteq \ker T(f|X_1)$. Such a vector field has a non-trivial flow line, on which f is constant, contradicting the finite-to-one condition. Thus $\dim X = \dim Y$.

Q. E. D.

(3.4) Let $f : \mathbb{R}^n \to \mathbb{R}^p$ be a polynomial mapping, and let $X \subseteq \mathbb{R}^n$ be a semialgebraic smooth submanifold. The set $Sg(f|X)$ of points in X where $f|X$ fails to be immersive is semialgebraic. Moreover, if $f|X$ is finite-to-one then $Sg(f|X)$ has dimension $< \dim X$.

Proof First, we show that TX is semialgebraic. For this we use the fact that a tangent line at a point on X is the limiting position of lines in $\mathbb{R}^n$ cutting X at two distinct points. We put

A = subset of $\mathbb{R}^n \times \mathbb{R}^n \times \mathbb{R}^n$ comprising triples (x, y, v) of collinear points with $x, y \in X$ and $x \neq y$.

B = subset of $\mathbb{R}^n \times \mathbb{R}^n \times \mathbb{R}^n$ comprising triples (x, y, v) of points with $x, y \in X$ and $x = y$.

Clearly, both A, B are semialgebraic. Let Δ be the diagonal in $\mathbb{R}^n \times \mathbb{R}^n$: then, under the natural identification of $T\mathbb{R}^n$ with $\Delta \times \mathbb{R}^n$ the semialgebraic set $\bar{A} \cap B$ will correspond to TX.

Now we show that $Sg(f|X)$ is semialgebraic. Observe first that we can suppose f a linear projection, (Factor f as the composite of the natural embedding in its graph, and a linear projection.) Let $S(\mathbb{R}^n)$ denote the unit sphere bundle in $T(\mathbb{R}^n)$. By the above the set $\ker Tf \cap TX \cap S(\mathbb{R}^n)$ is semialgebraic in $T(\mathbb{R}^n)$: its image under the projection $T(\mathbb{R}^n) \to \mathbb{R}^n$ is precisely $Sg(f|X)$, and semialgebraic, by the Tarski-Seidenberg Theorem.

Finally, assume $f|X$ finite-to-one. Suppose that $\dim Sg(f|X) = \dim X$, so there is an open set $U \subseteq X$ at no point of which f is immersive. We can suppose f has constant rank on U, replacing U by a smaller open set if necessary. It will now be possible to construct a non-trivial vector field

ξ on U with $\xi(U) \subseteq \ker Tf$. Such a vector field has a non-trivial flow line, on which f is constant - contradicting the hypothesis that f is finite-to-one. It follows that indeed $\dim Sg(f|X) < \dim X$.

Q. E. D.

The preliminary work is now complete and we are in a position to establish the existence of canonical partial stratifications for generic polynomial mappings.

(3.5) Let A, B be semialgebraic open sets in $\mathbb{R}^n$, $\mathbb{R}^p$ respectively, and $f : A \to B$ a polynomial mapping for which $f|\Sigma(f)$ is finite-to-one. f admits a canonical partial stratification $\mathcal{C}$ having only finitely many semialgebraic strata.

Proof Observe that the set Σ of critical points of f is semialgebraic, so that the set C of critical values is likewise semialgebriac, by the Tarski-Seidenberg Theorem. Put $c = \dim C$. We shall construct a filtration $C = C_c \supseteq C_{c-1} \supseteq \ldots\ldots$ of C by semialgebraic sets, closed in C , with $\dim C_j \leq j$, such that the following conditions are satisfied for each j with $M_j = C_j - C_{j-1}$.

(PS0) M_j is empty, or a smooth submanifold of P of dimension j over which each M_k is Whitney regular for $k > j$.

(PS1) $f^{-1}M_j \cap \Sigma$ is a smooth submanifold of N , and the restriction $f : f^{-1}M_j \cap \Sigma \to M_j$ is a local diffeomorphism.

(PS2) $f^{-1}M_k \cap \Sigma$ is Whitney regular over $f^{-1}M_j \cap \Sigma$ for $k > j$.

(PS3) $f^{-1}M_k - \Sigma$ is Whitney regular over $f^{-1}M_j \cap \Sigma$ for $k \geq j$.

Suppose inductively that $C_c, C_{c-1}, \ldots, C_j$ have been constructed according to this prescription. If $\dim C_j < j$ we take $C_{j-1} = C_j$ and are done. Suppose $\dim C_j = j$. We put

$$R_1 = \text{closure}\left\{\bigcup_{k>j} W(C_j , M_k)\right\}$$

which is semialgebraic, of dimension $< j$. Now put

$$C_j' = f^{-1}(C_j - R_1) \cap \Sigma$$

which is semialgebraic of dimension j , by (3.3) . We define

$$R_2 = R_2' \cup R_2''$$

where

$$R_2' = \text{closure}\left\{\bigcup_{k>j} W(C_j' , f^{-1}M_k \cap \Sigma)\right\}$$

and

$$R_2'' = \text{closure}\left\{\bigcup_{k>j} W(C_j' , f^{-1}M_k - \Sigma)\right\} .$$

Clearly R_2' , R_2'' (and hence R_2) are semialgebraic, of dimension $< j$. Next, put

$$R_3 = \text{closure of } Sg(f|C_j' - R_2)$$

which is semialgebraic, of dimension $< j$, by (3.4) . Also write

$$C_j'' = C_j' - R_2 - R_3$$

which is semialgebraic, of dimension j , and

$$R_4 = \text{closure of } W(C_j'', f^{-1}fC_j'' - \Sigma)$$

which is semialgebraic, of dimension $< j$. Finally, define

$$C_{j-1} = R_1 \cup f(R_2 \cup R_3 \cup R_4)$$

which is semialgebraic, of dimension $\leq (j - 1)$, by (3.2) .

That completes the induction step of the construction. We leave to the reader the task of formally checking that the partition of C by the M_j is the required canonical partial stratification of f .

Q. E. D.

Now let A, B be semialgebraic open sets in Euclidean spaces, and $f : A \to B$ a polynomial mapping which satisfies the Genericity Condition. By (3.5) f admits a canonical partial stratification $\mathcal{C}$. And by (3.1) f admits a Thom stratification $(\mathcal{S}, \mathcal{S}')$, namely the Thom stratification associated to $\mathcal{C}$: this we call a canonical Thom stratification for f . Summing up, we obtain the main result of Chapter I .

(3.6) Let A, B be semialgebraic open sets in Euclidean spaces, and let $f : A \to B$ be a polynomial mapping which satisifes the Genericity Condition : then f admits a canonical Thom stratification $(\mathcal{X}, \mathcal{X}')$.

It is this result which enables us to construct the crucial stratification of the jet space in Chapter IV . We conclude the present chapter by listing the basic properties of canonical partial stratifications required for Chapter IV. First, it is invariant under smooth equivalence of smooth mappings; precisely

(3.7) Let f_1, f_2 be smooth mappings for which there exist diffeomorphisms g, h with $f_1 \circ g = h \circ f_2$. Suppose that f_1 admits a canonical partial stratification $\mathcal{C}_1$, and let $\mathcal{C}_2$ be the stratification induced by h : then $\mathcal{C}_2$ is a canonical partial stratification for f_2 .

The proof consists of no more than a (tedious) formal checking - which is also the case for

(3.8) Let $f : N \to P$ be a smooth mapping which admits a canonical partial stratification $\mathcal{C}$. Let $P' \subseteq P$ be open, let $N' = f^{-1}P'$, and let $\mathcal{C}' = C|P'$: then $\mathcal{C}'$ is a canonical partial stratification for $f : N' \to P'$.

Another result for which it does not seem worthwhile writing down a proof is

(3.9) Let $f : N \to P$ be a smooth mapping which admits a canonical partial stratification $\mathcal{C}$. And let $U \subseteq N$ be an open set containing the critical set of f . Then $\mathcal{C}$ also provides a canonical partial stratification for $f|U$.

The final fact which we shall need to know later is

(3.10) For $1 \leqslant j \leqslant s$ let $f_j : N_j \to P$ be smooth mappings, with domains having the same dimension, and let $f : N \to P$ be their disjoint sum.

Suppose each f_j admits a Thom stratification $(\mathcal{S}_j, \mathcal{S}_j')$, and that the $\mathcal{S}_j'$ are in general position. f admits a Thom stratification $(\mathcal{S}, \mathcal{S}')$ with $\mathcal{S}' = \cap\, \mathcal{S}_j'$. Moreover, if the $(\mathcal{S}_j, \mathcal{S}_j')$ are all canonical then so too is $(\mathcal{S}, \mathcal{S}')$.

C H A P T E R I I

Stratifications and Flows

Klaus Wirtmüller

§1 Tubes

This chapter is concerned with the construction of continuous flows on stratified sets. Since strata are smooth manifolds we can clearly obtain a smooth flow on each stratum of a stratification by integrating a smooth vector field. But in general we cannot expect to obtain a continuous flow on the whole set by just putting the parts together. Therefore the vector fields we consider will be of a special nature: in a sense to be made precise they will be controlled along the boundary of each stratum in such a way that we do obtain continuous flows. Let us illustrate the idea by means of an example.

(1.1) Example. Stratify the plane $\mathbb{R}^2$ by $\{0, \mathbb{R}^2 - 0\}$ and let ξ be the vector field on $\mathbb{R}^2$ defined by

$$\xi(0) = 0, \quad \xi(x, y) = \frac{x \cdot \partial/\partial y - y \cdot \partial/\partial x}{\|(x, y)\|} \quad \text{for} \quad (x, y) \neq 0$$

(in standard coordinates). Clearly, the origin is a fixed point and all other flow lines are circles centred at the origin. Note that

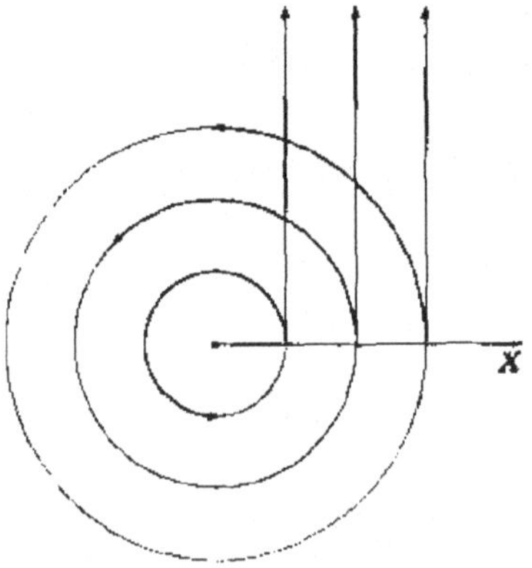

outside the origin all flow lines have constant speed $\|\xi\| = 1$ whence their angular velocities $\|\xi(x, y)\|/\|(x, y)\|$ tend to infinity as (x, y) approaches 0. Thus we have constructed a flow on $\mathbb{R}^2$ which is not differentiable at the origin. On the other hand distance to the origin is constant along any flow line, and this implies at once that the flow is continuous.

This simple type of control by the distance to the smaller stratum will not be sufficient to ensure continuity of the flow unless this stratum is a discrete set.

The next example indicates another type of control over a stratum of positive dimension.

(1.2) Example. Split $\mathbb{R}^3 = \mathbb{R} \times \mathbb{R}^2$ and stratify it by $\{\mathbb{R} \times 0, \ \mathbb{R} \times (\mathbb{R}^2 - 0)\}$. Then

$$\eta(t, x, y) = \partial/\partial t + \xi(x, y)$$

defines a vector field η on $\mathbb{R} \times \mathbb{R}^2$, with ξ as in (1.1). η induces a linear flow on the t-axis whereas the other flow lines spiral round the t-axis.

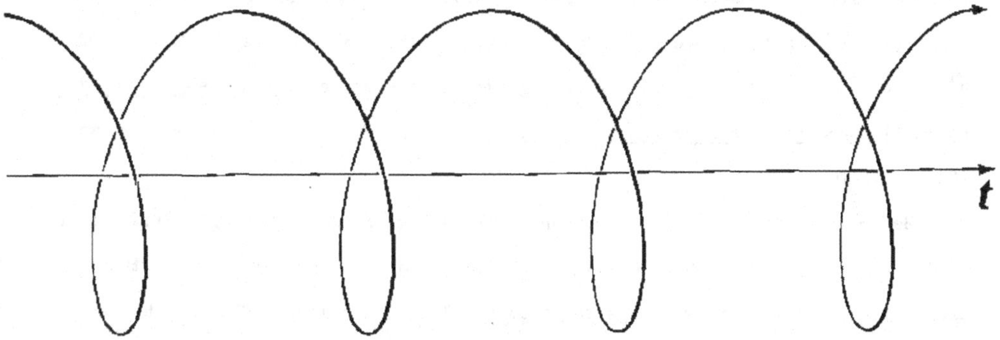

This time the proof that we have a continuous flow on $\mathbb{R}^3$ is based on two observations :

(1) the flow preserves distance to the t-axis, and

(2) under the canonical retraction $\mathbb{R} \times \mathbb{R}^2 \to \mathbb{R} \times 0$ the flow line through (t, x, y) (say) is mapped to the flow line through $(t, 0, 0)$ on the t-axis.

Our strategy in the general case will be to look for vector fields having properties (1) and (2) locally, i.e. in some neighbourhood of the "smaller" stratum. We will give a meaning to "distance" and "retraction" in this context by taking this neighbourhood to be a tube. But first of all let us introduce the following notion which will be convenient later on.

(1.3) Definition. Let X, Y, Z be subsets of a topological space N. We say that two maps g_1, g_2 defined on Z define the same germ at X (at (X, Y)) if there exist neighbourhoods U of X and V of Y in N such that g_1 and g_2 coincide on $Z \cap U$ (on $Z \cap U \cap V$, respectively).

Note that we do not require that X or Y be contained in Z. It should be clear that the composition of two suitable germs, or one germ and a map, is a well-defined germ.

Now we give the precise definition of "tube".

(1.4) Definition. Let $X \subseteq N$ be a submanifold of the smooth manifold N (X need not be closed in N). A tube at X is a quadruple $T = (E, \pi, \rho, e)$ where $\pi : E \to X$ is a (smooth) vector bundle, $\rho : E \to \mathbb{R}$ the quadratic function of a Riemannian metric on E, and $e : E \to N$ is the germ at $\zeta(X)$ of a local diffeomorphism, commuting with the zero section $\zeta : X \to E$ so that $e \circ \zeta$ is the germ (at X) of the inclusion $X \subseteq N$. Observe that $\pi^T = \pi \circ e^{-1} : N \to X$ and $\rho^T = \rho \circ e^{-1} : N \to \mathbb{R}$ are well defined germs at X; we call them the germs of the retraction and the distance function defined by T. By abuse of language we will often omit the superscript T.

Before we come to the construction of tubes let us look at Example (1.2) once more. If we let $g_0 : \mathbb{R} \times \mathbb{R}^2 \to \mathbb{R}$ be the projection to the first factor then property (2) means that g_0 maps the flow on $\mathbb{R}^3$ to the linear flow on the real line generated by the constant vector field $\partial/\partial t$. Conversely, given any submersion g retracting $\mathbb{R}^3$ onto $\mathbb{R}$ suppose we wish to construct a flow on $\mathbb{R}^3$, preserving the strata and mapping to the standard linear flow on $\mathbb{R}$ under g. For general g, no such flow will have the property (2) and the reason for this is that (2) refers to the standard tube rather than a tube which is compatible with g in the following sense.

(1.5) Definition. Let T be a tube as in (1.4) and let $g : N \to P$ be a map (or a germ of such a map at X). T is compatible with g if the germs (at X) g and $g \circ \pi^T$ are equal.

Thus each fibre of π^T should be contained in a fibre of g. We can expect to find compatible tubes only for very special g, and the remainder of this section is devoted to the proof of the following existence result.

(1.6) Theorem. Let N, $X \subseteq N$ and P be smooth manifolds, and let $g : N \to P$ be a smooth map germ at X such that $g|X$ is a submersion. Suppose $X_1 \subseteq X_0$ are (relatively) open subsets of X with $\bar{X}_1 \cap X \subseteq X_0$. If T_0 is a tube at X_0, compatible with g, then there exists a tube T at X, also compatible with g, such that $T|X_1 = T_0|X_1$.

Proof. Our proof depends on a standard method of constructing tubes, see [Lang] or [Bröcker - Jänich] : choose any spray ξ on N and let e_ξ be the germ at X of the associated exponential map, restricted to a normal bundle $\pi : E \to X$ of X in N. Then $T = (E, \pi, ?, e_\xi)$ is a tube at X.

We shall obtain the required extra properties of T from a careful choice of ξ and E. Let us deal with the latter first.

Let $Tg : TN \to TP$ denote the differential of g. Since $g|X$ is submersive the kernel rank of g is constant in a neighbourhood of X, and it is no loss of generality to assume that this neighbourhood is all of N. Then $\ker Tg$ is a subbundle of TN, and furthermore we have $\ker Tg + TX = TN$ over X.

Let $T_0 = (E_0, \pi_0, \rho_0, e_0)$. The differential of e_0 along the fibres of π_0 is a monomorphism

$$T^{\text{fibre}}(e_0) : E_0 \to TN|X_0$$

of vector bundles, by means of which we identify E_0 with its image in $TN|X_0$. Then we must have $E_0 \subseteq (\ker Tg)|X_0$ since T_0 is compatible with g.

Now if we pick a Riemannian metric on the bundle $TX_0 \cap (\ker Tg)|X_0 = \ker T(g|X_0)$ we also get a metric $\bar{\rho}_0$ on $(\ker Tg)|X_0$ splitting the latter bundle into the orthogonal sum of the Riemannian bundles $\ker T(g|X_0)$ and E_0 (here we think of ρ_0 as a Riemannian metric rather than a quadratic function). By a standard extension process we find a metric $\bar{\rho}$ on $(\ker Tg)|X$ such that $\bar{\rho} = \bar{\rho}_0$ over X_1. We let E be the orthogonal complement of $\ker(Tg|X)$ in $(\ker Tg)|X$ with respect to $\bar{\rho}$ and let ρ be $\bar{\rho}|E$. Then E is a normal bundle of X such that $E \subseteq (\ker Tg)|X$ and $E|X_1 = E_0|X_1$, i.e. for any choice of the spray ξ its solution curves will leave X in the right direction.

Thus $T = (E, \pi, \rho, e_\xi)$ will have all required properties if the spray ξ satisfies

(1) $\xi(\ker Tg) \subseteq T(\ker Tg)$

(2) $e_\xi | (E|X_1) = e_0 | (E|X_1)$.

But finding such a spray is a local problem in the following sense: if we can construct ξ locally near any point $x \in X$ then we find a locally finite covering of X by open subsets of N on which the local parts of ξ are defined. A partition of unity subordinate to this covering gives us a spray ξ on some neighbourhood of X, satisfying (1) and (2), and this is clearly enough.

We deal with the local problem of defining ξ near a point $x \in X$ by linearising the situation, as follows.

(a) let $x \in X_0$. Since $g|X_0$ is submersive we find charts for X_0 near x and for P near gx with respect to which $g|X_0$ is a linear projection. Choose linear coordinates in the normal space $E_0|x$ and a trivialisation of E_0 near x : via e_0 we obtain an N-chart near x extending the chart for X_0. Note that since T_0 is compatible with g, $g \circ e_0$ is constant along the fibres, so g is automatically linear in the N- and P-charts. Now let ξ, in the chart on N, be the standard geodesic spray on Euclidean space. We leave it to the reader to check that ξ satisfies (1) and (2).

(b) if $x \in X - X_0$ we define ξ as in case (a) but use any chart on N which linearises g. Then ξ clearly has property (1), and if we shrink to a neighbourhood of x staying well away from X_1 we may ignore (2). □

Note that Theorem (1.6) still holds if we allow the components of P to have different dimensions.

(1.7) Corollary. Theorem (1.6) remains true if g is not a germ at X but only at an open subset X_g of X containing $X - X_1$.

Proof. Choose an open subset N_g of N such that $X_g = X \cap N_g$ and apply Theorem (1.6) to N_g in place of N. We obtain a tube at X_g which coincides with T_0 over $X_1 \cap X_g$, hence a tube at $X_1 \cup X_g = X$. □

§2 Tube Systems

We return to the problem of controlling vector fields on a stratified set by means of tubes. Let us first study the cross-ratio example mentioned in the introduction in more detail.

(2.1) **Example.** Let $\gamma : \mathbb{R} \to (0, \infty)$ be a smooth function and let $A \subseteq \mathbb{R}^3$ be defined by the equation

$$(*) \qquad xy(x + y)(x - \gamma(t) \cdot y) = 0 .$$

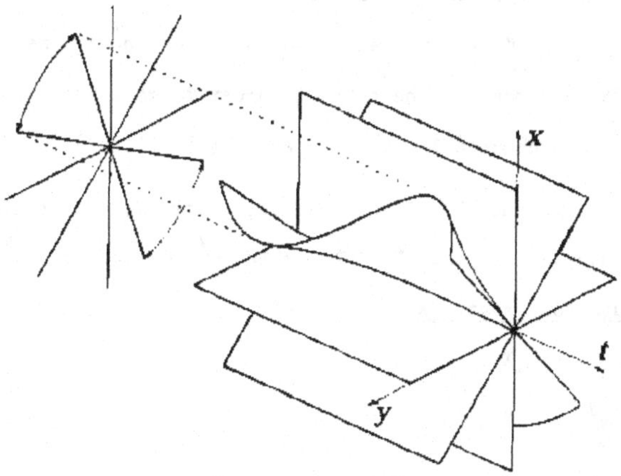

A is stratified by the t-axis X and its complement Y (in fact this is a canonical Whitney stratification). Adding $\mathbb{R}^3 - A$ as a third stratum, we obtain a Whitney stratification of $\mathbb{R}^3$. Suppose now we are looking for flows on $\mathbb{R}^3$ which

(1) preserve the strata, and

(2) project to the standard linear flow on the t-axis under the retraction $(t, x, y) \to t$.

First notice that this problem cannot have a smooth solution. For at every point $(t, 0, 0)$ on the t-axis, the limiting positions of tangent spaces $T_x A$ $(x \in A, x \to (t, 0, 0))$ consist of four planes in $\mathbb{R}^3$, each containing the t-axis. The cross-ratio defined by the corresponding concurrent lines in the projective plane is, after suitable ordering of the lines, just $\gamma(t)$. A smooth flow satisfying (1) and (2) must preserve the cross-ratios, which is impossible unless γ happens to be constant. Note that in fact we have shown that any strata-preserving C^1 flow must be trivial on the t-axis.

Next let us try to piece together a continuous solution, smooth on each stratum. Trivially condition (2) determines our flow on X. If we take T^X to be the canonical tube at X then we clearly find a unique solution on A which is controlled by T^X (i.e. commutes with the retraction and preserves the distance function of T^X). But for an arbitrary tube T^Y at Y it will generally be impossible to extend this solution further to a flow on $\mathbb{R}^3$ which is controlled by both T^X and T^Y. The obstruction comes from the fact that π^Y will in general retract neither along the fibres of π^X nor along the surfaces ρ^X = constant.

The conditions we have to impose on the tubes in order to make such extensions possible are most conveniently phrased in terms of germs (see (1.3)).

(2.2) <u>Definition</u>. Let X and Y be submanifolds of the smooth manifold N, and let T^X and T^Y be tubes at X and Y. We write π^X instead of π^{T^X} etc. Let us introduce the <u>commutation relations</u>

$$(\mathrm{CR}\pi) \qquad \pi^X \circ \pi^Y = \pi^X$$

$$(\mathrm{CR}\rho) \qquad \rho^X \circ \pi^Y = \rho^X ,$$

both being equations between germs at (X, Y). Furthermore, let Y' be a submanifold of another smooth manifold N', with a tube $T^{Y'}$, and let $f : N \to N'$ map Y into Y'. Then we will also have to consider

$$(\mathrm{CR}f) \qquad f \circ \pi^Y = \pi^{Y'} \circ f \quad \text{(as germs at } Y\text{)}.$$

The reason for introducing this last relation will become clear shortly. Note that the three commutation relations, put in a different way, require that T^Y be compatible with π^X, ρ^X and $\pi^{Y'} \circ f$ respectively. In view of the submersiveness condition in the existence theorem for tubes (1.6) the following lemma is the key to further progress.

(2.3) <u>Lemma</u>. <u>Let</u> X <u>and</u> Y <u>be disjoint smooth submanifolds of the smooth manifold</u> N, <u>and let</u> T^X <u>be a tube at</u> X. <u>If</u> Y <u>is Whitney regular over</u> X <u>then the map germ</u> (<u>at</u> X)

$$(\pi^X, \rho^X)|Y \; : \; Y \to X \times \mathbb{R}$$

is submersive.

Proof. If it is not, we find a sequence (y_i) in Y converging to some point $x \in X$ such that $(\pi^X, \rho^X)|Y$ is not submersive at y_i. Choosing a local trivialisation of the tube T^X near x we may assume that T^X is the standard tube at $X = \mathbb{R}^p \times 0$ in $N = \mathbb{R}^n = \mathbb{R}^p \times \mathbb{R}^{n-p}$. Since $(\pi^X, \rho^X) : \mathbb{R}^n \to \mathbb{R}^p \times \mathbb{R}$ is a submersion at y_i and $(\pi^X, \rho^X)|Y$ is not we conclude that $T_{y_i} Y$ is not transverse to the kernel of (π^X, ρ^X) in y_i. This kernel is the orthogonal complement of $\mathbb{R}^p \times 0 + \vec{xy_i}$ in $\mathbb{R}^n$. By compactness of the Grassmannians of subspaces of $\mathbb{R}^n$ we may assume $(T_{y_i} Y) \to T$ and $(\vec{xy_i}) \to L$ where T is a p-dimensional subspace and L a line in $\mathbb{R}^n$. Using a simple convergence lemma we see that T is not transverse to the orthogonal complement of $\mathbb{R}^p \times 0 + L$ in $\mathbb{R}^n$. But this implies $T_x X + L = \mathbb{R}^p \times 0 + L \not\subseteq T$, so Y is not Whitney regular over X at x. □

The strategy outlined so far will enable us to lift a smooth flow on a manifold to a controlled flow on a stratified set. It is natural (and will in fact become necessary) to look also at a more general problem: given a stratified map $f : A \to A'$ (say) and a controlled flow on A', does there exist a controlled flow on A which is mapped to the given one under f? If we restrict attention to two strata X and Y in A then there is nothing new about the situation provided f maps X and Y into a single stratum in A'. But if $f(X) \subseteq X'$ and $f(Y) \subseteq Y'$ (say) where X' and Y' are different strata then the requirement that the flow on A lift that on A' will interfere with the control exercised by the distance function ρ^X. A simple example is provided by (1.2) and (1.1) together with the projection $f : \mathbb{R} \times \mathbb{R}^2 \to \mathbb{R}^2$.

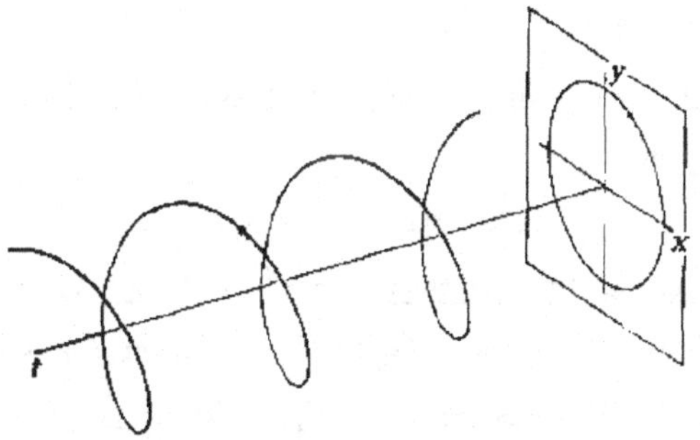

The example suggests that the control by ρ^X should be disposed of in these instances since a similar control is already effected by the map f together with the flow in the target: f keeps the spiralling flow line automatically at constant distance to the t-axis. On the other hand we do need control by the retraction π^X, and this is why (CRf) was introduced. We are heading now for a result analogous to (2.3) but involving π^X and f rather than π^X and ρ^X. Before we can state it we need a bit of technical preparation.

Let $f : N \to N'$ be a smooth map sending the smooth submanifolds X and Y in N submersively into X' and Y' respectively, where X' and Y' are smooth submanifolds of N'. Furthermore, suppose we are given tubes at X, X' and Y' such that (CRπ) holds for the pair (X', Y') and (CRf) holds for X and X'. Replacing the germ (at X') $\pi^{X'} : Y' \to X'$ by a smooth representative we form the pullback

$$\begin{array}{ccc} X \times_{X'} Y' & \xrightarrow{\text{proj}} & Y' \\ \downarrow{\scriptstyle \text{proj}} & & \downarrow{\scriptstyle \pi^{X'}} \\ X & \xrightarrow[f]{} & X' \end{array}$$

(i.e. $X \times_{X'} Y' = \{(x, y') \in X \times Y' : fx = \pi^{X'} y'\}$). Since $f : X \to X'$ is a submersion $X \times_{X'} Y'$ is a smooth manifold, and the commutation relations imply that the germ of (X, Y)

$$(\pi^X, \pi^{Y'} \circ f) : N \to X \times Y'$$

actually maps into $X \times_{X'} Y'$.

Clearly, this property does not depend on the choice of a representative above, and we are now ready to state the following.

(2.4) Lemma. If, under the assumptions above, Y is Thom regular over X with respect to f then

$$(\pi^X, f)|Y : Y \to X \times_{X'} Y'$$

is a submersive germ at X (notice that $f = \pi^{Y'} \circ f$ on Y).

Proof. If not, we find a sequence (y_i) in Y converging to some point $x \in X$ such that $(\pi^X, f)|Y$ is not submersive at y_i. If we put $x_i = \pi^X y_i$, $y_i' = fy_i$ and

$x'_i = fx_i$ then the tangent space to $X \times_{X'} Y'$ at (x_i, y'_i) is $T_{x_i}X \times_{T_{x'_i}X'} T_{y'_i}Y'$. Using the fact that Tf sends $T_{y_i}Y$ <u>onto</u> $T_{y'_i}Y'$, a simple calculation shows that $(\pi^X, f)|Y$ fails to be submersive at y_i if and only if the linear map $T_{y_i}\pi^X : \ker T_{y_i}(f|Y) \to \ker T_{x_i}(f|X)$ is not onto : the relevant maps are displayed in the diagram

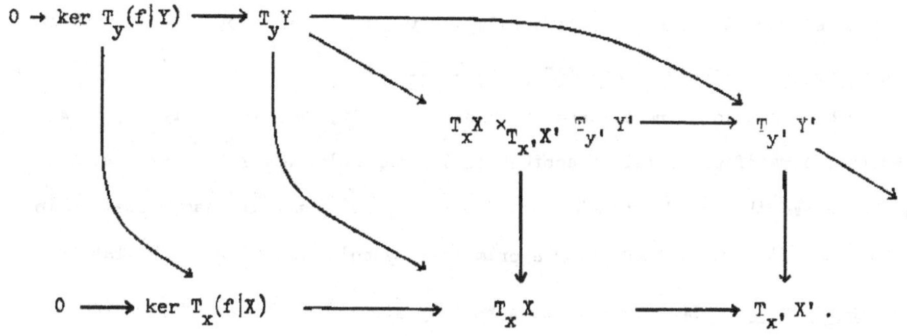

Passing to a subsequence we may assume that all spaces $T_{y_i}\pi^X[\ker T_{y_i}(f|Y)]$ have the same dimension and converge to a proper subspace S of T_xX, while the sequence $(\ker T_{y_i}(f|Y))$ itself converges to $T \subseteq T_xN$, say. Necessarily $(\ker T_{x_i}(f|X))$ converges to $\ker T_x(f|X)$, and we obtain the proper inclusion $T_x\pi^X(T) = S \subset \ker T_x(f|X)$. Since $T_x\pi^X$ is a retraction whose image contains $\ker T_x(f|X)$ we cannot have $\ker T_x(f|X) \subseteq T$, so Y is not Thom regular over X at x. □

The main result of the section will be that we can (under suitable conditions) assign a tube to each stratum so that all relevant commutation relations are satisfied. Let us first fix some notation to which we will stick throughout this and the next two sections.

Let $(A, \mathcal{A})$ be a Whitney stratified subset of the smooth manifold N, and let $a = \dim A$ (i.e. the maximal dimension of a stratum in $\mathcal{A}$). For each $i = 0, 1, \ldots, a$ we denote the union of all i-dimensional strata in $\mathcal{A}$ by M^i. By (I.1.1) each M^i is a smooth submanifold of N of dimension i. Notice also that the M^i are locally finite unions of strata, so a commutation relation holds for a pair (M^i, M^j) if and only if it holds for all pairs of strata (X, Y) with $X \subseteq M^i$, $Y \subseteq M^j$. It will in fact turn out to be more convenient to work with the manifolds M^i rather than with single strata whenever possible. Next we put, for $-1 \leq \alpha < \beta \leq a$,

$$A_\alpha^\beta = \bigcup_{i=\alpha+1}^{\beta} M^i \quad \text{and} \quad A^\beta = A_{-1}^\beta .$$

Note for later use that each A^β is a relatively closed subset of A and that the A_α^β are locally closed in A, again by (I.1.1). Notice that the A^β provide the filtration by dimension already used in Chapter I. If, more generally, $(A', \mathcal{A}')$ is a Whitney stratified subset of another smooth manifold N' such that $(\mathcal{A}, \mathcal{A}')$ stratifies a smooth map $f : N \to N'$ then let us agree to use the same notation in source and target but distinguish by a prime all symbols referring to the latter.

(2.5) Definition. Let $(A, \mathcal{A})$ be a Whitney stratified subset of the smooth manifold N. A tube system for $\mathcal{A}$ consists of one tube $T^i = (E^i, \pi^i, \rho^i, e^i)$ at each M^i, for $i = 0, 1, \ldots, a$. We say that the tube system is weakly controlled if all commutation relations of type (CRπ) hold; we call it controlled if furthermore all (CRρ) are satisfied. Let $f : N \to N'$ be smooth and let $(\mathcal{A}, \mathcal{A}')$ be a stratification of $f : A \to A'$. If $(T'^k)_{k=0}^{a'}$ is a tube system for A' then a tube system $(T^i)_{i=0}^{a}$ for A is controlled over (T'^k) if the T^i satisfy all relations of type (CRπ) and (CRf) and if (CRρ) holds for those pairs (X, Y) of strata in $\mathcal{A}$ with $fX \cup fY \subseteq M'^k$ for some k.

Note that in general "controlled over (T'^k)" does not imply "controlled" (unless, e.g. all strata of $\mathcal{A}'$ have the same dimension).

We can now state the result announced above.

(2.6) Theorem. Let N and N' be smooth manifolds, $f : N \to N'$ a smooth map. Let $A \subseteq N$, $A' \subseteq N'$, and suppose $(\mathcal{A}, \mathcal{A}')$ is a Thom stratification for $f : A \to A'$. Then for each weakly controlled tube system (T'^k) for $\mathcal{A}'$ there exists a tube system (T^i) for $\mathcal{A}$ which is controlled over (T'^k).

By putting N' = point we obtain

(2.7) Corollary. Every Whitney stratification admits a controlled tube system. □

Proof of (2.6). We proceed by double induction as follows: assume we have already constructed a tube system (T^i) for A^{a-1} which is controlled over (T'^k) (there is nothing to show for $a = 0$). By Theorem (1.6) we find a tube T^a at M^a satisfying condition (CRf) (with respect to the tubes T'^k in N'). In order to complete the proof we have to modify this tube near A^{a-1} so as to satisfy the various commutation relations involving strata in A of dimension less than a. We do this by downward induction on $\alpha < a$: assume the tube system induced on A^a_α is controlled over (T'^k) (for $\alpha = a - 1$ we have $A^a_\alpha = M^a$). If we can arrange that (T^i) restricted to $A^a_{\alpha-1}$ is also controlled over (T'^k) then the proof will be complete, for $\alpha = 0$ gives the result.

Therefore let us assume the tube system induced on A^a_α is controlled over (T'^k). First of all, we choose representatives for the germs e^i, for $\alpha \leq i < a$: thus each e^i is a diffeomorphism of some neighbourhood of the zero section in E^i onto a neighbourhood of M^i in N, and we denote the latter by $|T^i|$. Note that we may consider π^i and ρ^i as maps defined on $|T^i|$. Let for each i with $\alpha \leq i < a$,

$$Q^i = M^a \cap |T^i| .$$

We partition Q^i according to where points in Q^i and their projections to M^i are mapped by f. So let, for all $0 \leq k,\ell \leq a'$, $Q^i(k,\ell)$ be the subset of all points $y \in Q^i$ such that f maps y into M'^ℓ and $\pi^i y$ into M'^k. It will turn out that (CRπ) and (CRf) are relevant over $Q^i(k, \ell)$ for $k < \ell$ whereas

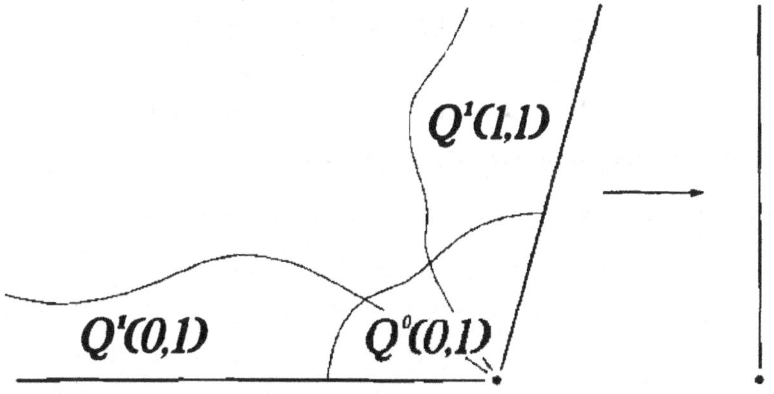

over $Q^i(k, k)$ we have to look at (CRπ) and (CRρ). We therefore set

$$Q^i_\rho = \bigcup_{k=0}^{a'} Q^i(k, k) .$$

We obtain a partition

$$Q^i = Q^i_\rho \cup \bigcup_{k \neq \ell} Q^i(k, \ell)$$

of Q^i into a finite number of open subsets : openness follows at once from the fact that f maps strata into strata.

Now we choose representatives for the tubes T'^k too. We can shrink the domains of all diffeomorphisms e^i and e'^k so that the following statements become true.

(1) as germs at Q^i, $\pi^i \circ \pi^a = \pi^i$ for $\alpha < i < a$ (possible by downward induction hypothesis)

(2) as germs at $Q^\alpha \cap Q^i$, $\pi^\alpha \circ \pi^i = \pi^\alpha$ for $\alpha < i < a$

(3) as germs at $Q^\alpha_\rho \cap Q^i$, $\rho^\alpha \circ \pi^i = \rho^\alpha$ for $\alpha < i < a$ ((2) and (3) by upward induction hypothesis)

(4) as germs at $M'^\ell \cap |T'^k|$, $\pi'^k \circ \pi'^\ell = \pi'^k$ for $0 \leqslant k < \ell \leqslant a'$ (since (T'^k) is weakly controlled)

(5) $Q^i(k, \ell) = \emptyset$ if $k > \ell$, for all i (in view of (I.1.1)

(6) if $X^p \in A$ and $X'^q \in A'$ are strata such that $fX \subseteq X'$ then f maps $|T^p|X|$ into $|T'^q|X'|$ (by continuity of f)

(7) as germs at $Q^i(k, \ell)$, $f \circ \pi^i = \pi'^k \circ f$ for $\alpha \leqslant i < a$ and $0 \leqslant k \leqslant \ell \leqslant a'$ (again by induction hypothesis)

(8) $(\pi^\alpha, \rho^\alpha)|Q^\alpha_\rho : Q^\alpha_\rho \to M^\alpha \times \mathbb{R}$ is a submersion (by Whitney regularity and Lemma (2.3))

(9) $(\pi^\alpha, f)|Q^\alpha(k, \ell) : Q^\alpha(k, \ell) \to M^\alpha \times_{M'^k} (M'^\ell \cap |T'^k|)$ is a submersion for $0 \leqslant k < \ell \leqslant a'$ (by Thom regularity, (7) and Lemma (2.4)).

Now we apply Corollary (1.7) with

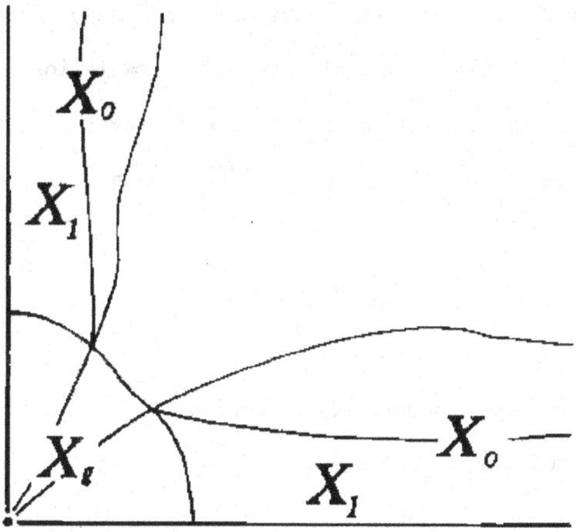

the following data:

$$X = \bigcup_{\alpha \leq i < a} Q^i, \quad X_o = \bigcup_{\alpha < i < a} Q^i, \quad T_o = T^a | X_o, \quad X_g = Q^\alpha .$$

Thus we have to specify g as a germ at $Q^\alpha = Q^\alpha_\rho \cup \bigcup_{k<\ell} Q^\alpha(k, \ell)$: we set

$$g = (\pi^\alpha, \rho^\alpha) \quad : \quad N \to M^\alpha \times \mathbb{R} \qquad \text{near } Q^\alpha_\rho$$

$$g = (\pi^\alpha, \pi'^\ell \circ f) : N \to M^\alpha \times_{M'^k} (M'^\ell \cap |T'^k|) \quad \text{near } Q^\alpha(k, \ell).$$

Then $g|X_g$ is a submersion by (8) and (9), and $T_o|X_g$ is compatible with g; for over each Q^i, $\alpha < i < a$, we have

$$\pi^\alpha \circ \pi^a \underset{(2)}{=} \pi^\alpha \circ \pi^i \circ \pi^a \underset{(1)}{=} \pi^\alpha \circ \pi^i \underset{(2)}{=} \pi^\alpha$$

$$\rho^\alpha \circ \pi^a \underset{(3)}{=} \rho^\alpha \circ \pi^i \circ \pi^a \underset{(1)}{=} \rho^\alpha \circ \pi^i \underset{(3)}{=} \rho^\alpha ,$$

and T^a is compatible with f by assumption. Finally, to get a subset $X_1 \subseteq X$ to which (1.7) can be applied we shrink the tubes T^i $(\alpha < i < a)$ a bit further, obtain open subsets Q^i_1 in Q^i and set $X_1 = \bigcup_{\alpha < i < a} Q^i_1$. Since all spaces involved are normal and X_o is a neighbourhood of the relatively closed subset $X - X_g$ in X we can arrange $X - X_g \subseteq X_1 \subseteq \bar{X}_1 \cap X \subseteq X_o$.

Applying (1.7) we obtain a new tube T at X (with retraction π, say) which extends $T^a|X_1$ and is compatible with g. T satisfies all required commutation relations: this follows from the very construction of g everywhere but over Q^α_ρ, where we have to verify (CRf). But $Q^\alpha_\rho = \bigcup Q^\alpha(k, k)$, and over $Q^\alpha(k, k)$ we compute

$$f \circ \pi = \pi'^k \circ f \circ \pi \underset{(7)}{=} f \circ \pi^\alpha \circ \pi = f \circ \pi^\alpha \underset{(7)}{=} \pi'^k \circ f .$$

We complete the induction step by (shrinking and) extending T over all M^a. This is done by another application of Theorem (1.6), this time with $g = f$. We let the resulting tube at M^a be the new T^a. □

§3 Vector Fields

We shall now give precise definitions for the class of vector fields we propose to integrate.

(3.1) Definition. Let $(A, \mathcal{A})$ be a stratified subset of the smooth manifold N. A stratified vector field ξ on A is a collection of smooth vector fields ξ^i on M^i, for $i = 0, 1, \ldots, a$. By abuse of language, we also denote by ξ the (in general discontinuous) vector field on A with values in the tangent space of N, which equals ξ^i on M^i. If (T^i) is a tube system for $\mathcal{A}$ then the control conditions for ξ at a stratum $Y \in \mathcal{A}$ are

$$(\text{VF}\pi) \qquad T\pi^Y \circ \xi = \xi \circ \pi^Y$$

$$(\text{VF}\rho) \qquad T\rho^Y \circ \xi = 0 \;;$$

both are equations between germs at Y. We call ξ weakly controlled if (VFπ) holds for all $Y \in \mathcal{A}$, and controlled if furthermore (VFρ) holds for all $Y \in \mathcal{A}$. Next let $f : A \to A' \subseteq N'$ be a smooth map, stratified by $(\mathcal{A}, \mathcal{A}')$, and let ξ' be a stratified vector field on A'. We say that the stratified vector field ξ on A lifts ξ' if the condition

$$(\text{VF}f) \qquad Tf \circ \xi = \xi' \circ f$$

holds on A. Instead of (VFρ) we also consider the following weaker condition on

ξ at a stratum $Y \in \mathcal{A}$: if f maps Y into $Y' \in \mathcal{A}$ then, as a germ at Y,

$$(\mathrm{VF}\rho f) \qquad T\rho^Y \circ \xi \mid f^{-1}Y' = 0 .$$

We say that ξ is <u>controlled over ξ'</u> if ξ satisfies all conditions of type (VFf), (VFπ) and (VFρf).

The reader may verify himself that the vector fields in (1.1) and (1. 2) are controlled in the appropriate sense. The following existence result shows that the notions just defined are wide enough to be useful.

(3. 2) <u>Theorem</u>. <u>Let</u> $f : N \to N'$ <u>be a smooth map between manifolds, let</u> $A \subseteq N$ <u>and</u> $A' \subseteq N'$ <u>be subsets, and let</u> $(\mathcal{A}, \mathcal{A}')$ <u>be a Thom stratification for</u> $f : A \to A'$. <u>Let</u> (T^i) <u>and</u> (T'^k) <u>be tube systems for</u> $\mathcal{A}$ <u>and</u> $\mathcal{A}'$, <u>and suppose that</u> (T^i) <u>is controlled over</u> (T'^k). <u>Then any weakly controlled vector field</u> ξ' <u>on</u> A' <u>admits a lift (i.e. a stratified vector field</u> ξ <u>on</u> A <u>lifting it) which is controlled over</u> ξ'.

<u>Proof</u>. We proceed by induction on the dimension a of A. So let us assume we have already defined ξ on A^{a-1}. We choose representatives for all tubes T^i and T'^k as in the proof of (2. 6) such that the following hold (using the same notation):

(1) $\pi^i \circ \pi^j = \pi^i$ on $Q^i \cap Q^j$, for $0 \leq i < j < a$

(2) $\rho^i \circ \pi^j = \rho^i$ on $Q^i_\rho \cap Q^j$, for $0 \leq i < j < a$

(3) $Q^i(k, \ell) = \emptyset$ if $k > \ell$, $0 \leq i < a$

(4) if $X^p \in \mathcal{A}$, $X'^q \in \mathcal{A}'$, $fX \subseteq X'$ then $f(|T^p|X|) \subseteq |T'^q|X'|$

(5) $f \circ \pi^i = \pi'^k \circ f$ on $Q^i(k, \ell)$, for $0 \leq i < a$, $0 \leq k \leq \ell \leq a'$

(6) $(\pi^i, \rho^i) : Q^i_\rho \to M^i \times \mathbb{R}$ is a submersion, for $0 \leq i < a$

(7) $(\pi^i, f) : Q^i(k, \ell) \to M^i \times_{M'^k} (M'^\ell \cap |T'^k|)$ is a submersion, for $0 \leq i < a$, $0 \leq k < \ell \leq a'$

(8) $T\pi^i \circ \xi = \xi \circ \pi^i$ on $A^{a-1} \cap |T^i|$, for $0 \leq i \leq a - 2$

(9) $T\rho^i \circ \xi \quad (y) = 0$ if $y \in A^{a-1} \cap |T^i|$ and $fy = f\pi^i y$

(10) $T\pi'^k \circ \xi' = \xi' \circ \pi'^k$ on $A' \cap |T'^k|$

(11) $Tf \circ \xi = \xi' \circ f$ on A^{a-1}.

By slightly shrinking the tubes T^i we also obtain subsets $F^i \subseteq Q^i$ which are closed in M^a; of course (1) to (11) still hold with F^i in place of Q^i (we put $F^i(k, \ell) = F^i \cap Q^i(k, \ell)$ and $F^i_\rho = F^i \cap Q^i_\rho$). The proof will be complete if we can construct a smooth vector field ξ^a on M^a which satisfies

$$\text{(VF}\pi\text{)} \qquad T\pi^i \circ \xi^a = \xi^i \circ \pi^i \quad \text{on } F^i$$

$$\text{(VF}\rho f\text{)} \qquad T\rho^i \circ \xi^a = 0 \quad \text{on } F^i_\rho$$

$$\text{(VFf)} \qquad Tf \circ \xi^a = \xi' \circ f \quad \text{on } M^a$$

for all $i = 0, \ldots, a-1$ (notice that π^i and ρ^i are now maps rather than map germs). Clearly these conditions are local and define a convex subset of the linear space of all smooth vector fields on M^a, so we may define ξ^a locally and glue the pieces together by means of a partition of unity on M^a. Let $y \in M^a$: there are three cases to consider.

(a) $y \notin \bigcup_{o \leq i < a} F^i$: since f maps strats into strata it sends an open neighbourhood of y in M^a into some M'^{ℓ}, and since it does so submersively $f|M^a$ is, near y, smoothly equivalent to a linear projection of vector spaces. Thus we can find a vector field ξ^a satisfying (VFf) on a neighbourhood of y in M^a, and we choose this neighbourhood disjoint from $\bigcup_{0 \leq i < a} F^i$. Then (VF$\pi$) and (VF$\rho$f) hold trivially.

(b) $y \in F^j$ for some j, $o \leq j < a$. We choose j maximal with this property. Then either $y \in F^j_\rho$ or $y \in F^j(\ell, m)$ for some $o \leq \ell < m \leq a'$.

(b_1) $y \in F^j_\rho$: by (6), $(\pi^j, \rho^j) : Q^j_\rho \to M^j \times \mathbb{R}$ is a submersion. We choose ξ^a near y as a lift of the vector field $\xi^j \times 0$ under (π^j, ρ^j), arguing as in case (a). Then (VFπ) and (VFρf) hold for $i = j$ by construction, and all other conditions follow by computation :

$$Tf \circ \xi^a \underset{(5)}{=} Tf \circ T\pi^j \circ \xi^a = Tf \circ \xi^j \circ \pi^j \underset{(11)}{=} \xi' \circ f \circ \pi^j \underset{(5)}{=} \xi' \circ f$$

and for $i < j$

$$T\pi^i \circ \xi^a \underset{(1)}{=} T\pi^i \circ T\pi^j \circ \xi^a = T\pi^i \circ \xi^j \circ \pi^j \underset{(8)}{=} \xi^j \circ \pi^i \circ \pi^j \underset{(1)}{=} \xi^i \circ \pi^j$$

$$T\rho^i \circ \xi^a \underset{(2)}{=} T\rho^i \circ T\pi^j \circ \xi^a \underset{(8)}{=} T\rho^i \circ \xi^j \circ \pi^j \underset{(9)}{=} 0$$

(b_2) $y \in F^j(\ell, m)$, $\ell < m$. Here (10) and (11) imply that there is a well-defined vector field $\xi^j \times_{M'^\ell} \xi'^m$ on $M^j \times_{M'^\ell} (M'^m \cap |T'^\ell|)$, which we may (locally) lift to $Q^j(\ell, m)$, using (7). Then (VFf) and (VFπ), for $i = j$, hold by construction, and (VFπ) follows for $i < j$ as in (a). Finally condition (VFρf) is empty on $Q^j(\ell, m)$. □

(3.3) Corollary. Let $f : N \to P$ be a smooth map between manifolds, and let $(A, \mathcal{A})$ a Whitney stratified subset of N such that f maps each stratum of $\mathcal{A}$ submersively into P. Let (T^i) be a controlled tube system for $\mathcal{A}$. Then each smooth vector field on P can be lifted to a controlled vector field on A.

Proof. This is nothing but the special case $A' = N' = P$, $\mathcal{A}' = \{P\}$ of (3.2). The reader may check for himself that under the assumptions made $(\mathcal{A}, \{P\})$ is indeed a Thom stratification for $f : A \to P$. □

§4 Flows

The final step in our discussion will be the proof that controlled vector fields do define continuous flows. But let us first look at the ways in which an arbitrary stratified vector field may fail to do so.

(4.1) Example. Let ξ be the vector field on the plane $\mathbb{R}^2$ given by $\xi(x, 0) = \partial/\partial x$ and $\xi(x, y) = 0$ for $y \neq 0$, $(x, y) \in \mathbb{R}^2$. This is a stratified vector field in the obvious way. We can integrate it and obtain a flow $\mathbb{R} \times \mathbb{R}^2 \to \mathbb{R}^2$ which is discontinuous along the x-axis. A less trivial example is

(4.2) Example. Let ξ be defined on $\mathbb{R}^2$ as $\xi(x, 0) = \partial/\partial x$ and $\xi(x, y) = \partial/\partial x + \partial/\partial y$ for $y \neq 0$. Recall that we expect a flow on $\mathbb{R}^2$ to be defined at least on a neighbourhood of $0 \times \mathbb{R}^2$ in $\mathbb{R} \times \mathbb{R}^2$. But if we collect all maximal flow lines of ξ we obtain a mapping $D \to \mathbb{R}^2$ where the set D is obtained from $\mathbb{R} \times \mathbb{R}^2$ by deleting all points of the plane $\{t = y\}$ for which $t \neq 0$. Clearly the difficulty is that the flow lines through points just below

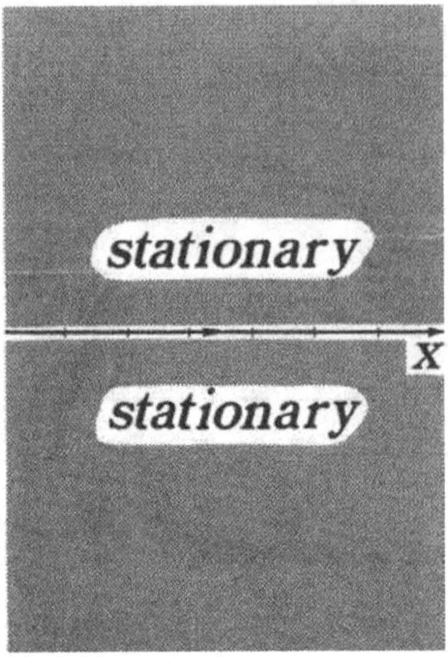

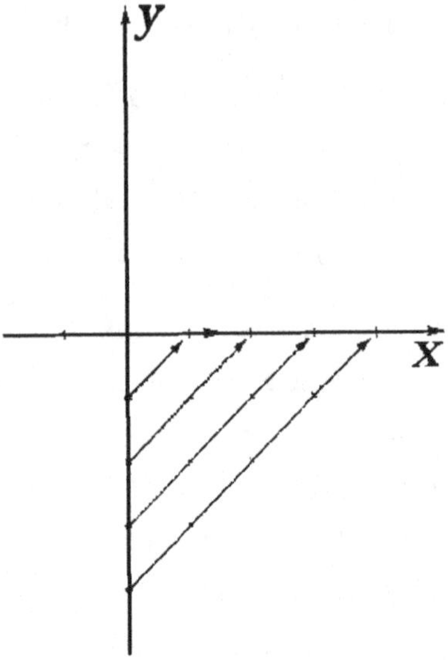

the x-axis are too short-lived as they approach the x-axis (in the picture we have marked time intervals of equal lengths on the flow lines).

We return to the general discussion: let ξ be a stratified vector field on A. Integrating each vector field ξ^i on M^i we obtain smooth flows $\theta^i : D^i \to M^i$ where D^i is the maximal domain belonging to ξ^i; D^i is an open subset of $\mathbb{R} \times M^i$ containing $0 \times M^i$. Of course θ^i is given by

$$\theta^i(0, x) = x, \quad \frac{d}{dt}\, \theta^i(t, x) = \xi(x).$$

For any $x \in A$, D^i intersects $\mathbb{R} \times \{x\}$ in an open interval $(t_x^-, t_x^+) \times x$, with $-\infty \leq t_x^- < 0 < t_x^+ \leq \infty$. We let $\theta_x : (t_x^-, t_x^+) \to A$ denote the flow line through x (sending t to $\theta(t, x)$). Now setting $D = \bigcup_{i=0}^{a} D^i$ we obtain a map $\theta : D \to A$ which has the algebraic properties of a flow on A though in general it will not be continuous (nor need D be open).

(4.3) Definition. The stratified vector field is locally integrable if D contains a neighbourhood of $0 \times A$ on which θ is continuous. If furthermore $D = \mathbb{R} \times A$ then ξ is called globally integrable.

The following lemma characterises the maximality of D.

(4.4) Lemma. Let ξ be locally integrable. Then D is open in $\mathbb{R} \times A$ and θ is continuous everywhere on D. Furthermore, if $x \in A$ and $t_x^+ < \infty$ then $\theta_x : [0, t_x^+) \to A$ is a proper map, and t_x^+ is unique with respect to this property (of course a similar statement is true for t_x^-).

Proof. The first part of the lemma is, for smooth flows, a standard result in differential topology, and the proof easily generalises (see [Lang] or [Bröcker-Jänich]). It is also well-known that under the assumption $t_x^+ < \infty$ the inverse image of a compact set $C \subseteq A$ under $\theta_x | [0, t_x^+)$ is actually contained in a compact subinterval of $[0, t_x^+)$, and so is itself compact. Conversely, if $t < t_x^+$ is a positive number then the set $\theta_x | [0, t)^{-1} (\theta_x[0, t]) = [0, t)$ is not compact, so $\theta_x | [0, t)$ cannot be proper. □

Note that the lemma applies in particular to the smooth vector fields ξ^i on A^i, so we have

(4.5) Corollary. If ξ is any stratified vector field on A then for any $x \in M^i$ such that $t_x^+ < \infty$ the map $\theta_x : [0, t_x^+) \to M^i$ is proper.

We are now ready to prove the main result.

(4.6) Theorem. Let N and N' be smooth manifolds and $f : N \to N'$ a smooth map. Let $A \subseteq N$ and $A' \subseteq N'$ be subsets, and $(\mathcal{A}, \mathcal{A}')$ a stratification for $f : A \to A'$. Let ξ and ξ' be stratified vector fields on A respectively A' such that ξ is controlled over ξ' with respect to some tube system (T^i) for $\mathcal{A}$. Suppose that the set A is locally closed in N. Then ξ is locally integrable if ξ' is.

Proof. By downward induction on $\alpha = a, a-1, \dots, 0$ we may assume that $\xi | A_\alpha^a$ is locally integrable. The inductive step consists in proving

(1) D contains a neighbourhood of $0 \times M^\alpha$ in $[0, \infty) \times A$,

(2) θ is continuous at all points of a neighbourhood of $0 \times M^\alpha$ in $\mathbb{R} \times M^\alpha$.

For we may replace $[0, \infty)$ in the first statement by $\mathbb{R}$ (look at $-\xi$), and then the induction hypothesis and (4.4) imply that $\theta | \bigcup_{i=\alpha+1}^{a} D^i$ is continuous, and since $\bigcup_{i=\alpha+1}^{a} D^i$ is open in D the result will follow.

For the proof of (1) it will be convenient to shorten $\Theta(J \times C)$ to JC if $J \times C$ is a subset of D. Now let $x \in M^{\alpha}$, $X \in A$ the stratum containing x, and C a compact neighbourhood of x in X. Since $\xi|X$ is a smooth vector field on X we find a number $t > 0$ such that $[0, t] \times C$ is contained in D. We choose a representative of the tube $T^{\alpha} = (E, \pi, \rho, e)$ with e mapping the total space of a small open disc bundle in E diffeomorphically onto an open set $|T| \subseteq N$, such that $T\pi \circ \xi = \xi \circ \pi$ on $|T|$ and $T\rho \circ \xi = 0$ on $f^{-1}X' \cap |T|$ ($X' \in A'$ is the stratum containing fX). We regard π and ρ as maps defined on $|T|$. For a sufficiently small positive number ϵ the "disc bundle" $E([0, t]C, \epsilon)$, defined as $A^a_{\alpha-1} \cap \pi^{-1}([0, t]C) \cap \rho^{-1}[0, \epsilon]$, is a <u>compact</u> subset of $|T|$, for $A^a_{\alpha-1}$, as a locally closed subset of N, is a locally compact space. More generally, we define

$$E(\cdot, \delta) = A^a_{\alpha-1} \cap \pi^{-1}(\cdot) \cap \rho^{-1}[0, \delta]$$
$$S(\cdot, \delta) = E(\cdot, \delta) \quad \cap \rho^{-1}(\delta)$$

where $0 < \delta \leq \epsilon$ and the dot stands for an arbitrary subset of $[0, t]C$. Making ϵ smaller, we may assume that $[0, t] \times f(E(C, \epsilon))$ is contained in D' (the maximal domain associated with ξ'), for since ξ lifts ξ' we have necessarily $[0, t] \times f(C) \subseteq D'$, and by (4.4), D' is open in $\mathbb{R} \times A'$. Since the compact set $S([0, t]C, \epsilon)$ is contained in A^a_{α} the inductive assumption guarantees that for some positive number $s \leq t$ the set $[-s, 0] \times S([0, t]C, \epsilon)$ is contained in D. But $\Theta([-s, 0] \times S([0, t]C, \epsilon))$ cannot meet C, and we find a positive number $\delta < \epsilon$ such that $\Theta([-s, 0] \times S([0, t]C, \epsilon)$ is disjoint from $E(C, \delta)$.

We show that $[0, s] \times E(C, \delta)$ is contained in D. So let $y \in E(C, \delta)$. There are three cases to consider.

(a) $y \in C$: this case is trivial.

(b) $fy \notin X'$: we want to show $s < t^+_y$, so let us assume $t^+_y \leq s$ and look at the set

$$J = (\theta_y|[0, t^+_y))^{-1}\{E([0, s]C, \epsilon) \cap f^{-1}(\theta'_{fy}[0, s])\} .$$

J is an open subset of $[0, t^+_y)$, for the control conditions imply $\pi \circ \theta_y = \theta_{\pi y}$ and $f \circ \theta_y = \theta'_{fy}$, and by construction $\theta_y|[0, t^+_y)$ avoids $S([0, s]C, \epsilon)$. On the other hand, as θ'_{fy} does not meet X', $E([0, s]C, \epsilon) \cap f^{-1}(\theta'_{fy}[0, s])$ is a compact subset of A^a_{α}, so the inductive hypothesis implies via (4.4) that

J is compact. Since $0 \in J$ we conclude that $J = [0, t_y^+)$, hence that $[0, t_y^+)$ is compact, which is absurd. Thus $s < t_y^+$ as desired. We are left with

(c) $fy \in X'$ but $y \notin C$, hence $\rho y > 0$: we argue as in (b), replacing $A_{\alpha-1}^a$ by $A_{\alpha-1}^a \cap |T|$, $(A', \mathcal{A}')$ by $[0, \infty)$, stratified by the origin and its complement, f by ρ and ξ' by the zero vector field.

Now the proof of (1) is complete. In order to prove assertion (2) we use the same construction in a slightly different way, as follows. Again let $x \in M^\alpha$. First of all, we choose C, t and ϵ so small that $[-t, t] \times E([-t, t]C, \epsilon)$ is contained in D. This is possible since D contains a neighbourhood of $(0, x)$ in $\mathbb{R} \times A$ by what we have already seen. We prove that Θ is continuous in (r, x) where r is any number in $(-t, t)$. Given a neighbourhood V of $\Theta(r, x)$ in A we make C and ϵ even smaller so that $E(RC, \epsilon)$ is contained in V for some interval $R \subseteq (-t, t)$ containing r. We find a $\delta > 0$ such that $\Theta([-t, t] \times S([-t, t]C, \epsilon))$ is disjoint from $E(C, \delta)$, and we show that Θ maps $R \times E(C, \delta)$ into V. By control through π it suffices to prove that Θ sends $[-t, t] \times E(C, \delta)$ into $E([-t, t]C, \epsilon)$. So let $y \in E(C, \delta)$: we may restrict ourselves to the case $fy \notin X$ as in (1) above. Look at

$$K = [-t, t] \cap \theta_y^{-1}\{E([-t, t]C, \epsilon) \cap f^{-1}(\theta'_{fy}[-t, t])\} .$$

It follows easily that K is open and closed in $[-t, t]$, and since $0 \in K$ we obtain $K = [-t, t]$. □

The case N' = point deserves a separate statement.

(4.7) Corollary. A controlled vector field on a locally compact stratified set is locally integrable. □

The hypothesis that A be locally compact (or, equivalently, locally closed in N) cannot be dropped: stratify the plane $\mathbb{R}^2$ (with coordinates x and y) by the x-axis and its complement. By deleting all points of the y-axis but the origin we obtain a Whitney stratified set on which the vector field $\partial/\partial x$ is controlled but not locally integrable.

We conclude the section with the following observation:

(4.8) Lemma. Let $f : N \to N'$ be a smooth map, $A \subseteq N$, $A \subseteq N'$ and $(\mathcal{A}, \mathcal{A}')$ a stratification for $f : A \to A'$. Assume that for each stratum $X \in \mathcal{A}$ the restriction $f : \bar{X} \cap A \to A'$ is a proper map. Further let ξ and ξ' be stratified vector fields on A and A' such that ξ lifts ξ' and assume that ξ is locally integrable. Then if ξ' is globally integrable then ξ is, too.

(4.9) Corollary. The conclusion holds in particular for all f such that $f : A \to A'$ is proper. □

Proof of (4.8) If not, we find a point $x \in A$ such that $(t_x^-, t_x^+) \neq R$, and we may assume $t_x^+ < \infty$. By (4.4), $\theta_x | [0, t_x^+)$ is proper. If $X \in \mathcal{A}$ is the stratum containing x then θ_x actually maps into X, and $\theta_x : [0, t_x^+) \to \bar{X} \cap A$ is also proper, hence the composition $\theta'_{fx} = f \circ \theta_x : [0, t_x^+) \to A'$ is proper. By (4.4) again, $t_{fx}^+ = t_x^+ < \infty$, so ξ' cannot be globally integrable. □

§5 Applications

(5.1) Definition. Let A be a subset of the smooth manifold N, $\mathcal{A}$ a stratification of A, and f a smooth map from N into another smooth manifold P. We say that the stratified set $(A, \mathcal{A})$ is (topologically) trivial over P if there are a stratified set $(F, \mathcal{F})$ and a homeomorphism $h : A \approx P \times F$ with $f = pr_P \circ h$, sending $\mathcal{A}$ to the product stratification $P \times \mathcal{F}$ (pr_P denotes the projection to P). We call $(A, \mathcal{A})$ locally trivial over P if each point $y \in P$ has an open neighbourhood V such that $(A \cap f^{-1}V, \mathcal{A} \cap f^{-1}V)$ is trivial over V. The homeomorphisms $h : A \cap f^{-1}V \approx V \times F$ are called local trivialisations.

The following result gives a sufficient condition for local triviality over P.

(5.2) Theorem. Let $(A, \mathcal{A})$ be a locally closed Whitney stratified subset of the smooth manifold N, and let $f : N \to P$ be a smooth map such that for each stratum $X \in \mathcal{A}$, $f|X$ is a submersion and $f|\bar{X} \cap A$ is a proper map. Then $(A, \mathcal{A})$ is locally trivial over P.

Proof. The assertion being local with respect to P, we may assume that $P = \mathbb{R}^p$. We apply (2.6), (3.3), (4.6) and (4.8): each globally integrable smooth vector field on $\mathbb{R}^p$ admits a globally integrable stratified vector field on A lifting it. In particular, let $\xi_1, \ldots, \xi_p$ be lifts of the constant coordinate vector fields $\partial/\partial t_1, \ldots, \partial/\partial t_p$ on $\mathbb{R}^p$, and denote the flows generated on A by $\theta_1, \ldots, \theta_p$. If we let $(F, \mathcal{F})$ be the fibre $A \cap f^{-1}(0)$ with its induced Whitney stratification then a trivialisation of $(A, \mathcal{A})$ over $\mathbb{R}^p$ is given by

$$\mathbb{R}^p \times (A \cap f^{-1}(0)) \to A$$

$$(t_1, \ldots, t_p, x) \to \theta_1(t_1, \theta_2(t_2, \ldots, \theta_p(t_p, x))). \quad \square$$

Theorem (5.2) is known as Thom's First Isotopy Lemma, for in the special case $A = N$, $P = \mathbb{R}$ the proof we have given consists in constructing an isotopy of N moving each stratum of $\mathcal{A} \cap f^{-1}(0)$ to one in $\mathcal{A} \cap f^{-1}(1)$. Note that we cannot expect to find <u>smooth</u> local trivialisations as we have already seen in Example (2.1).

The next result gives us a great deal of information on the local structure of Whitney stratifications.

(5.3) <u>Definition.</u> Let $(A, \mathcal{A})$ be a stratified subset of the smooth manifold N. Let x be a point in A and $X \in \mathcal{A}$ the stratum containing x. We say that $\mathcal{A}$ is <u>(topologically) locally trivial at x</u> if there exist a stratified set $(F, \mathcal{F})$ and, for some $y \in F$, a local homeomorphism $h : A \approx X \times F$ sending x to (x, y) such that the one-point set $\{y\}$ is a stratum of $\mathcal{F}$ and h sends $\mathcal{A}$ to the product stratification $X \times \mathcal{F}$. $\mathcal{A}$ is <u>locally trivial</u> if it is so at every point in A.

(5.4) <u>Theorem.</u> <u>Let</u> $(A, \mathcal{A})$ <u>be a locally closed Whitney stratified subset of the smooth manifold</u> N, <u>and let</u> X <u>be a stratum in</u> $\mathcal{A}$. <u>Then there exist a smooth sphere bundle</u> $\phi : M \to X$, <u>a closed Whitney stratified subset</u> $(S, \mathcal{S})$ <u>of</u> M <u>and a neighbourhood</u> Z <u>of</u> X <u>in</u> N <u>such that</u>

(1) $(S, \mathcal{S})$ is locally trivial over X (with respect to ϕ)

(2) there exists a homeomorphism from the pair $(Z, A \cap Z)$ onto the pair of mapping cylinders $(Z_\phi, Z_{\phi|S})$ which is the identity on X (recall that Z_ϕ is obtained from $M \times [0, 1] + X$ by identifying $(y, 0) \in M \times [0, 1]$ with $\phi y \in X$).

Proof. Choose a tube $T = (E, \pi, \rho, e)$ at X and pick a representative for e mapping the closed ϵ-disc bundle $\{y \in E : \rho y \leq \epsilon(\pi y)\} \subseteq E$ diffeomorphically onto a neighbourhood $|T|$ of X in N (here ϵ is a suitable smooth positive function on X). By making ϵ small enough we may suppose that

(1) $A \cap |T|$ is closed in $|T|$ (for A is locally closed in N)

(2) for any stratum $Y \in \mathcal{A}$, $Y \neq X$, the map $(\pi, \rho) : Y \cap |T|^{\circ} \to X \times \mathbb{R}$ is a submersion (by (2.4) and local finiteness of $\mathcal{A}$).

Replacing ρ by $\frac{1}{2}\ \frac{\rho}{\epsilon \circ \pi}$ normalises $\epsilon \equiv 2$, and then the map

$$(\pi, \rho) : |T|^{\circ} - X \to X \times (0, 2)$$

is proper. $|T|^{\circ} - X$ is Whitney stratified by the open set $|T|^{\circ} - A$ and strata of the form $Y \cap |T|^{\circ}$, $Y \in \mathcal{A}$, and we may apply Theorem (5.2) to $|T|^{\circ} - X$ and (π, ρ). But, recalling the proof of (5.2), we restrict the choice of local coordinates in $X \times (0, 2)$ for this application: we allow only charts $\Omega : \mathbb{R}^{\dim X} \times \mathbb{R} \to X \times (0, 2)$ of the form

$$(x, t) \to (\omega(x), \frac{2e^{t}}{1 + e^{t}})$$

where $\omega : \mathbb{R}^{\dim X} \to X$ is a local chart for X. We define $Z \subseteq |T|$ and $M \subseteq |T|$ by $\rho \leq 1$ and $\rho = 1$ respectively, and let $\phi = \pi|M$ and $(S, \mathcal{S}) = (A \cap M, \mathcal{A} \cap M)$ (notice that M intersects each stratum in $\mathcal{A}$ transversely). On application of (5.2), or rather its proof, we obtain

(1) $(S, \mathcal{S})$ is locally trivial over X (with respect to ϕ): just ignore the vector field $\partial/\partial t$, and

(2) there is a homeomorphism $h : M \times (0, 1] \to Z - X$, making

$$\begin{array}{ccc} M \times (0, 1] & \xrightarrow{\ h\ } & Z - X \\ & {\scriptstyle \phi \times id} \searrow \quad \swarrow {\scriptstyle (\pi, \rho)} & \\ & X \times (0, 1] & \end{array}$$

commutative and mapping $S \times (0, 1]$ to $(A \cap Z) - X$: lift and integrate $\partial/\partial t$ only, obtain a flow θ on $|T|^{\circ} - X$, and let

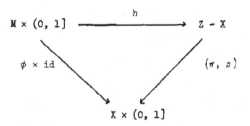

$$h(y, r) = \theta(\log \frac{r}{2-r}, y).$$

From the diagram in (2) it follows at once that h extends to a homeomorphism $Z_\phi \approx Z$ having the required properties. □

Notice incidentally that Z_ϕ is embedded in the vector bundle E by $(x, t) \to t \cdot e^{-1}x$; $(x, t) \in M \times [0, 1]$, and is Whitney stratified by $\{X\} \cup \mathcal{S} \times (0, 1]$ and that we have in fact constructed a homeomorphism between Z and Z_ϕ preserving stratifications. From this observation and the fact that, for (good) topological spaces U, V, W and a map $f : U \to V$, $Z_{id_W \times f}$ is naturally homeomorphic to $W \times Z_f$, we obtain at once the following.

(5.5) <u>Corollary</u>. <u>Let</u> N <u>be a smooth manifold and</u> $A \subseteq N$ <u>a locally closed subset. Then any Whitney stratification of</u> A <u>is topologically locally trivial.</u> □

Again, we cannot expect to find smooth local trivialisations: (2.1) provides a counterexample. On the other hand, topological local triviality is sufficient for the following application:

(5.6) <u>Theorem</u>. <u>Let</u> $(A, \mathcal{A})$ <u>be a locally closed Whitney stratified subset of the smooth manifold</u> N. <u>Then the partition</u> $\mathcal{A}^c$ <u>of</u> A <u>into the connected components of all strata in</u> $\mathcal{A}$ <u>is a Whitney stratification of</u> A.

<u>Proof</u>. We only have to check the Local Finiteness Condition, and we do this by induction on the dimension of the ambient space N, using the conclusion (and the notation) of Theorem (5.4) as follows: we may apply the inductive hypothesis to the subset $S \subseteq M$ (for $\dim M = \dim N - 1$). Thus if $x \in X$ and C is a compact neighbourhood of x in X then $S \cap \phi^{-1}C$ meets only a finite number of strata in $\mathcal{S}^c$. But since $A \cap Z$ is homeomorphic to $Z_{\phi|S}$ by a homeomorphism preserving strata each stratum of $\mathcal{A}^c$ meeting $A \cap Z \cap \pi^{-1}C$ must also intersect $S \cap \phi^{-1}C$, and of course different strata in $\mathcal{A}^c$ intersect $S \cap \phi^{-1}C$ in different strata of $\mathcal{S}^c$. Thus $A \cap Z \cap \pi^{-1}C$ is a neighbourhood of x in A meeting only finitely many strata of $\mathcal{A}^c$. □

(5.7) <u>Corollary</u>. $\mathcal{A}^c$ <u>satisfies the Frontier Condition.</u>

Proof. Let X and Y be strata in $\mathcal{A}^{\circ}$. $X \cap \bar{Y}$ is clearly closed in X. By local triviality it is also open in X. Since X is connected we have either $X \cap \bar{Y} = \phi$ or $X \subseteq \bar{Y}$. □

The result needed in Chapter IV is an analogue of (5.2) for stratified mappings rather than stratified sets. It is well-known as Thom's Second Isotopy Lemma.

(5.8) Theorem: Let

$$N \xrightarrow{F} N' \xrightarrow{f} P$$

be a diagram of smooth manifolds and maps, let $A \subseteq N$ and $A' \subseteq N'$ be locally closed subsets, and let $(\mathcal{A}, \mathcal{A}')$ be a Thom stratification for $F : A \to A'$ such that f maps each stratum of $\mathcal{A}'$ submersively into P. Suppose furthermore that for all strata $X \in \mathcal{A}$ and $X' \in \mathcal{A}'$, the restrictions $F : \bar{X} \cap A \to A'$ and $f : \bar{X}' \cap A' \to P$ are proper maps. Then the stratified map $(F : A \to A', \mathcal{A}, \mathcal{A}')$ is (topologically) locally trivial over P, i.e. for every point $y \in P$ there exist an open neighbourhood V of y in P and homeomorphisms

$$h : V \times F^{-1}f^{-1}(y) \cap A \to F^{-1}f^{-1}(V) \cap A$$

$$h' : V \times f^{-1}(y) \cap A' \to f^{-1}(V) \cap A' ,$$

preserving the natural stratifications, such that

$$\begin{array}{ccccc}
V \times F^{-1}f^{-1}(y) \cap A & \xrightarrow{\ \mathrm{id} \times F\ } & V \times f^{-1}(y) \cap A' & \xrightarrow{\ pr_V\ } & V \\
\downarrow{\scriptstyle h} & & \downarrow{\scriptstyle h'} & & \| \\
F^{-1}f^{-1}(V) \cap A & \xrightarrow{\ F\ } & f^{-1}(V) \cap A' & \xrightarrow{\ f\ } & V
\end{array}$$

commutes.

Proof. As in (5.2), we may assume $P = \mathbb{R}^p\ (=V)$. Again, consider the coordinate vector fields on $\mathbb{R}^p$, lift them to A', and lift them further to A, using (2.6) and (3.2). By (4.6) and (4.8) we obtain a flow on A covering those on A' and P. We define the trivialisations h and h' as in (5.2). □

Now a standard property of fibre bundles implies the following.

(5.9) <u>Corollary.</u> <u>If, under the assumptions of (5.8),</u> P <u>is connected then the topological type of the map</u>

$$F_y \;:\; F^{-1}f^{-1}(y) \cap A \;\to\; f^{-1}(y) \cap A'$$

<u>given by restricting</u> F <u>to the fibres over</u> y <u>is independent of</u> $y \in P$.

<u>Proof.</u> Partition P according to the topological type of F_y, $y \in P$. By (5.8) this is a partition of P into open subsets, hence it is the trivial partition. □

The Second Isotopy Lemma can easily be generalised to the case of a sequence of smooth maps

$$N_\ell \xrightarrow{F_\ell} N_{\ell-1} \to \dots \to N_1 \xrightarrow{F_1} N_0 \xrightarrow{f} P .$$

Details are left to the reader.

CHAPTER III

Unfoldings of smooth map-germs

by

Andrew du Plessis
(U.C.N.W., Bangor).

§0. Introduction

Throughout this chapter, all manifolds, mappings etc. will be taken to be smooth.

The mathematics described here arises from the study of smooth stability of mappings, and in particular the associated local theory.

For mappings, smooth stability is easily defined: a mapping is smoothly stable if all mappings sufficiently close to it (with respect to the Whitney topology) are equivalent to it under the natural action of diffeomorphisms of the source and target manifolds.

For map-germs, however, the immediate 'local' version of this (which, to make sense, would have to use as equivalence relation the action of diffeomorphisms preserving source and target points) is not the 'right' one (in the sense that one would hope that all germs defined by a stable map should be stable germs) - for, with respect to this definition, only germs of maximal rank would be stable (for clearly such germs form an open equivalence class whose complement is not a neighbourhood of any of its points, and so a fortiori contains no equivalence classes which are neighbourhoods of any of their points). Moreover, the topology for the space of map-germs implicitly being used here (via comparison of all derivatives at the source point) is apparently too coarse - it is, of course, completely appropriate for analytic map-germs, but it does not separate, for example, the germs $(\mathbb{R}, 0) \to (\mathbb{R}, 0)$ defined by e^{-x^2} and 0, any of whose representatives will always be separated by the Whitney topology.

The difficulties, of course, arise from our dancing too close attendance on the source and target points. Thus to match more closely the stability theory for mappings, we seem forced to consider representatives of germs: essentially, for 'nearby' germs, we need to consider representatives defined on 'nearby' open sets. This looks a very difficult theory to work with ...

The concept of unfolding (originally due to Thom) permits a very neat finesse of these difficulties. In the simplest version, one considers germs of finite-dimensional parametrized families of germs as a germ

$$F : (N \times U,\ x_0 \times u_0) \to (P \times U,\ y_0 \times u_0)$$

which preserves u-levels (i.e. representatives $\tilde{F}$ of F have the property that $\tilde{F}(N \times \{u\}) \subset P \times \{u\}$ for u in some neighbourhood of u_0 in U). We call such a family an unfolding of f, where $f : (N, x_0) \to (P, y_0)$ is the map-germ such that $(f(x), u_0) = F(x, u_0)$. If the germ f is smoothly stable, we expect any such unfolding to be trivial up to equivalence (that is, there should exist level-preserving germs of diffeomorphism H of $(N \times U,\ x_0 \times u_0)$, K of $(P \times U,\ y_0 \times u_0)$ such that $K \circ F \circ H^{-1} = f \times 1_U$).

Of course it is not quite clear that this notion of stability is equivalent to stability as it might be introduced via 'uniform representatives' (if we had formalized that notion properly!). However, it does stand in the 'right' relation to stability of mappings (see, for example, Guillemin and Golubitsky [Go]), and we will take this as justification for considering stability of map-germs in this way).

It turns out that stability in this sense has a very neat algebraic equivalent ('infinitesimal stability') leading to a very elegant classification of stable map-germs (which was first obtained by Mather; in fact, in this chapter we will meet a large proportion of the beautiful mathematics he introduced. in [MaIII], [MaIV] [MaV], to do this).

However, it is not entirely satisfactory to restrict attention to stable mappings, since they are not always generic (that is, there exist (compact) manifolds N, P such that the set of stable mappings $N \to P$ is not dense in the space of all mappings $N \to P$ (with respect to the Whitney topology); see [MaV]). It will, therefore, be the aim of this chapter to describe an extension to the notion of stability for map-germs which is almost as easily handled algebraically and whose 'globalization' is generic.

Let $f : (N, x_0) \to (P, y_0)$ be a map-germ. If f is not stable, so that germs 'nearby' f are not all in the same equivalence class, we might nevertheless hope that all 'nearby' germs are 'contained' (up to equivalence) in some unfolding

$F : (N \times U, x_0 \times u_0) \to (P \times U, y_0 \times u_0)$. More precisely we hope that for any unfolding $G : (N \times U', x_0 \times u_0') \to (P \times U', y_0 \times u_0')$ of f there exist germs Φ, Ψ such that the following diagram commutes:

$$\begin{array}{ccc} (N \times U, x_0 \times u_0) & \xrightarrow{F} & (P \times U, y_0 \times u_0) \\ \uparrow \Phi & & \uparrow \Psi \\ (N \times U', x_0 \times u_0') & \xrightarrow{G} & (P \times U', y_0 \times u_0') \end{array}$$

where Φ, Ψ have the form

$$\Phi(x,u') = (\phi_{u'}(x), \lambda(u')) \ , \ \Psi(y,u') = (\psi_{u'}(y), \lambda(u'))$$

with $\phi_{u_0'} = 1_N$, $\psi_{u_0'} = 1_P$ and λ a map-germ $(U',u_0') \to (U,u_0)$; that is, F is <u>versal</u> in a category of 'parametrized unfoldings' of f (whose morphisms (the pair (Φ,Ψ)) we have described above).

Sadly, the condition that f should have a versal parametrized unfolding, while still being fairly easily computable, does not always have generic 'globalization' either (counter-examples are described by Mather in [MaII]; but see also, for example, the work of Wasserman [Wa] where, following unpublished notes of Mather, the case where f is a function is extensively studied. In this case, since stability itself is generic for functions, the condition is quite satisfactory; and indeed very much stronger genericity statements can be made (see Looijenga [Lo])).

Thus we need to think a little harder. What we shall do is to remove the dependence on the parameter space. With this in view, let us begin the formal mathematics of this chapter:

(0.1) <u>Definition</u>

Let $f : (N,x_0) \to (P,y_0)$ be a map-germ.

An <u>unfolding</u> of f is a triple (F,i,j) of map-germs, where i,j are immersions and j is transverse to F , such that the diagram

$$\begin{array}{ccc} (N',x_0') & \xrightarrow{F} & (P',y_0') \\ \uparrow i & & \uparrow j \\ (N,x_0) & \xrightarrow{f} & (P,y_0) \end{array}$$

is cartesian (that is, the diagram is commutative, and the mapping (i,f) of N into the submanifold $\{(x',y) \in N' \times P | F(x') = j(y)\}$ is a diffeomorphism onto).

The dimension of (F,i,j) as an unfolding of F is $\dim.N' - \dim.N$.

We may choose co-ordinates for this situation as follows: let $(y_1,\dots,y_p,u_1,\dots,u_k)$ be a co-ordinate system for (P',y_0') such that $\mathrm{Im}(j)$ is given by $u_1 = \dots = u_k = 0$. Of course, then, $(y_1 \circ j,\dots,y_p \circ j)$ is a co-ordinate system for (P,y_0). Moreover, since j is transverse to F, $\{u_1 \circ F,\dots,u_k \circ F\}$ extends to a co-ordinate system $(x_1,\dots,x_n,u_1 \circ F,\dots,u_k \circ F)$ for (N',x_0'). Then, since the diagram is cartesian, $\mathrm{Im}(i)$ is defined by $u_1 \circ F = \dots = u_k \circ F = 0$, so that $(x_1 \circ i,\dots,x_n \circ i)$ is a co-ordinate system for (N,x_0).

Clearly, with respect to these co-ordinates, F 'preserves u-levels'; so that this notion of unfolding is just a 'co-ordinate-free' version of that introduced previously. From this point of view it is worth observing that essentially what we are doing here is choosing a submersion $\pi : (P',y_0') \to (U,u_0)$ such that $\pi^{-1}(u_0) = \mathrm{Im}(j)$ - one choice amongst many possibilities, whereas no choice is given us in the parametrized version.

(0.2) Definition

An unfolding (F,i,j) is trivial if there exist retractions r and s to i and j respectively such that $f \circ r = s \circ F$

(Recall that $r : (N',x_0') \to (N,x_0)$ is a retraction to i if $r \circ i = 1_N$; similarly, of course, $s \circ j = 1_P$)

With respect to co-ordinate systems as chosen previously, we may identify $N' = N \times U$, $P' = P \times U$. Then $R = (r,1_N)$ is a level-preserving diffeomorphism of $(N \times U, x_0 \times u_0)$, $S = (s,1_P)$ is a level-preserving diffeomorphism of $(P \times U, y_0 \times u_0)$, and $(f \times 1_N) \circ R = S \circ F$. So this notion of triviality is precisely the 'co-ordinate-free' version of that introduced earlier.

(0.3) <u>Definition</u>

A map-germ $F : (N,x_0) \to (P,y_0)$ is <u>stable</u> if any unfolding of F is trivial.

Thus far, then, there is no difference between this co-ordinate-free approach and the theory suggested earlier. However, in the case that f is not stable, asking that all 'nearby' germs to a germ f are 'contained' (up to equivalence) in some unfolding of f is again presumably to seek a 'versal' unfolding, but this time in a different category, in which the objects (unfoldings of f) are essentially the same as before, but in which there are many more morphisms.

(0.4) <u>Definition</u>

(a) A morphism

$$(\phi,\psi) : (F,i,j) \to (F',i',j')$$

of unfoldings of f is a commutative diagram of map-germs:

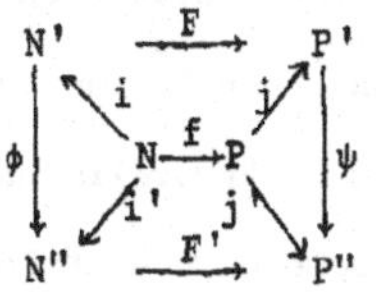

(b) An unfolding (F',i',j') of f is <u>versal</u> if for every unfolding (F,i,j) of f there exists a morphism

$$(F,i,j) \to (F',i',j') .$$

Let us observe, then, that a map-germ f is stable if and only if it is a versal unfolding of itself (the retractions (r,s) of (0.2) giving the morphism $(F,i,j) \to (f,1_N,1_P)$). In fact this statement generalizes as follows: an unfolding (F,i,j) of a map-germ f is versal if and only if the map-germ F is stable (when we call F a stable unfolding). This seems to be a very deep fact (certainly, I am not aware of any straightforward geometric approach to it), more or less the key to the classification of stable map-germs, and the proof of it therefore yields many other interesting results. In fact, it is these subsidiary results which are the useful ones for the purposes of this book (indeed, we could have avoided the concept of versality altogether, though with consequent more or less complete loss of motivation); these useful facts are:-

(1) A stable (equivalently, versal) unfolding is 'essentially' unique (for details, see §3).

(2) A stable unfolding is, with respect to suitable co-ordinates, a germ of polynomial mapping (see §5).

(3) The equivalence class, as a germ, of a stable unfolding is determined by the contact class of the germ it unfolds (we will introduce the notion of contact-equivalence in §4); moreover, a stable unfolding satisfies an important transversality condition relative to the action of the contact group in the jet-space. (see §6).

(4) The property of having a stable unfolding is 'generic' (in a sense which will be explained in §7).

Proofs of these facts (and of the equivalence of versality and stability for unfoldings) are based upon some rather subtle manipulations of modules of vector fields over rings of function-germs. The preliminaries to the algebra are presented in §1. For the moment, some of the style of argument we will be using can be understood from the lemma which follows, which will be useful in the succeeding sections.

However, just before stating and proving this lemma, it seems worthwhile to state explicitly that this chapter is mostly expository rather than original. Many of the concepts are due originally to Thom, while most of the deep methods were developed by Mather. I am happy to have this opportunity of expounding their profound ideas. I am also grateful to my co-authors, in particular E.J.N. Looijenga, and also to C.T.C. Wall, for much help with the material presented here.

(0.5) <u>Definition</u>

Let θ_{N,x_0} be the set of germs of vector fields on (N,x_0) (we shall write θ_N for this where no confusion will arise).

(0.6) Lemma

Let (F,i,j) be an unfolding for the map-germ $f : (N,x_0) \to (P,y_0)$, and let $\pi : (P',y_0') \to (U,u_0)$ be a (germ of) submersion such that $\pi^{-1}(u_0) = \mathrm{Im}(j)$.

Then (F,i,j) is trivial if and only if for any $\tau \in \theta_U$ there exist $\eta \in \theta_{P'}$, $\xi \in \theta_{N'}$ such that

$$\begin{cases} \eta \text{ is a lift for } \tau \text{ over } \pi \quad (\text{i.e. } T\pi \circ \eta = \tau \circ \pi) \\ \xi \text{ is a lift for } \eta \text{ over } F \quad (\text{i.e. } TF \circ \xi = \eta \circ F) \end{cases}$$

Proof

'Only if': If (F,i,j) is trivial, let r,s be the retractions as in (0.2). Suppose $\tau \in \theta_u$.

It is clear that $(s,\pi) : (P',y_0') \to (P \times U,\ y_0 \times u_0)$ is a germ of diffeomorphism, so that there exists a unique $\eta \in \theta_{P'}$ such that $T(s,\pi) \circ \eta = \tau \circ p_P$ (Here $p_P : P \times U \to U$ is the natural projection, and we identify $\tau \circ p_P$ as a (germ of) vector field on $P \times U$ via the natural splitting $T(P \times U) = TP \oplus TU)$. In particular, then, η is a lift for τ over π .

Similarly, $(r,\pi \circ F) : (N',x_0') \to (N \times U,\ x_0 \times u_0)$ is a germ of diffeomorphism at x_0' , so that there exists a unique $\xi \in \theta_{N'}$ such that $T(r,\pi \circ F) \circ \xi = \tau \circ p_N$

Thus

$$T(f \circ r,\pi \circ F) \circ \xi = T(f \times 1_U) \circ (\tau \circ p_N) = (\tau \circ p_P) \circ F = T(s,\pi) \circ \eta \circ F$$

But $(f \circ r,\pi \circ F) = (s,\pi) \circ F$ (from the definition of r,s), and so

$$T(s,\pi) \circ TF \circ \xi = T(s,\pi) \circ \eta \circ F .$$

Since (s,π) is a germ of diffeomorphism, we have, therefore, that $TF \circ \xi = \eta \circ F$, so that ξ is a lift for η over F .

'If': Let $(u_1,\dots,u_k)$ be a system of local co-ordinates at (U,u_0) , and let η_i be a lift for $\frac{\partial}{\partial u_i}$ over π , ξ_i a lift for η_i over F , for each $i = 1,\dots,k$.

Let $\psi_i : (P' \times \mathbb{R},\ y_0' \times 0) \to (P',y_0')$, $\phi_i : (N' \times \mathbb{R},\ x_0' \times 0) \to (N',x_0')$ be the germs of local flow for η_i,ξ_i respectively (for facts about integration of vector fields, see Chapter II). Then we have

$$u_k \circ \pi \circ \psi_{i,t} = u_k \circ \pi + t.\delta_{ik} \ , \ F \circ \phi_{i,t} = \psi_{i,t} \circ F$$

(where $\psi_{i,t}(y') = \psi_i(y',t)$, $\phi_{i,t}(x') = \phi_i(x',t)$).

Now define $r : (N',x_0') \to (N,x_0)$, $s : (P',y_0') \to (P,y_0)$ by

$$\begin{cases} r(x') = i^{-1} \circ \phi_{k,-v_k} \circ \ldots \circ \phi_{1,-v_1}(x') & (\text{where } v_i = u_i \circ \pi \circ F(x')\ ) \\ s(y') = j^{-1} \circ \psi_{k,-w_k} \circ \ldots \circ \psi_{1,-w_1}(y') & (\text{where } w_i = u_i \circ \pi(y')\ ) \end{cases}$$

The use of i^{-1},j^{-1} makes sense: recall that $Im(j) = \pi^{-1}(u_0)$, $Im(i) = (\pi \circ F)^{-1}(u_0)$. And it is also clear that r,s are respectively retractions to i,j .

Now

$$f \circ r(x') = f \circ i^{-1} \circ \phi_{k,-v_k} \circ \ldots \circ \phi_{1,-v_1}(x') = j^{-1} \circ F \circ \phi_{k,-v_k} \circ \ldots \circ \phi_{1,-v_1}(x')$$

$$= j^{-1} \circ \psi_{k,-v_k} \circ \ldots \circ \psi_{1,-v_1} \circ F(x') = s \circ F(x')$$

(since $w_i(F(x')) = v_i(x')$)

and so $f \circ r = s \circ F$, as required: thus (F,i,j) is trivial.

§1. Preliminaries, mostly algebraic

(1.1) **Definition**

The local ring C_{N,x_0} is the ring of function-germs $(N,x_0) \to \mathbb{R}$ (with addition and multiplication defined pointwise).

This ring has a maximal ideal m_{N,x_0} consisting of all germs $(N,x_0) \to (\mathbb{R},0)$.

We shall write C_N, m_N for C_{N,x_0}, m_{N,x_0} when no confusion can arise.

(1.2) **Lemma**

Let $(x_1,\ldots,x_n)$ be a system of local co-ordinates for (N,x_0) . Then $x_1,\ldots,x_n$ represent a C_N-basis for m_N .

Proof

Let ϕ represent an element of m_N . Then

$$\begin{aligned}\phi(x) &= \phi(x) - \phi(0)\\ &= \sum_{i=1}^{n} \{\phi(0,\ldots,0,x_i,\ldots,x_n) - \phi(0,\ldots,0,x_{i+1},\ldots,x_n)\}\\ &= \sum_{i=1}^{n} \int_0^1 \frac{\partial}{\partial t} \{\phi(0,\ldots,0,tx_i,x_{i+1},\ldots,x_n)\}.dt\\ &= \sum_{i=1}^{n} x_i \int_0^1 \frac{\partial\phi}{\partial x_i} (0,\ldots,0,tx_i,x_{i+1},\ldots,x_n)\}.dt\end{aligned}$$

Thus the germ of ϕ is indeed a C_N-linear combination of the germs of the x_i's .

Further, by taking differentials (giving a C_N-linear homomorphism $m_N \to T^*N_{x_0}$) we see at once that the x_i's must be linearly independent (since the dx_i's are); and so they do indeed form a basis.

(1.3) **Definition**

Let $f : (N,x_0) \to (P,y_0)$ be a map-germ. Then a ring homomorphism $f^* : C_P \to C_N$ is defined by $f^*(\phi) = \phi \circ f$.

Note that $f^*m_P \subset m_N$.

Now let us consider a situation which has already arisen in our study of unfoldings:

(1.4) <u>Lemma</u>

Let

$$(U,u_0) \xrightarrow{i} (V,v_0) \xrightarrow{\pi} (W,w_0)$$

be a sequence of germs such that i is an immersion and π a submersion with $\pi^{-1}(w_0) = \mathrm{Im}(i)$.

Then the sequence

$$0 \to \pi^*(m_W).C_V \subset C_V \xrightarrow{i^*} C_U \to 0$$

is exact and (non-canonically) split.

<u>Proof</u>

Let $(w_1,\ldots,w_e)$ be a system of local co-ordinates for (W,w_0) and let $(v_1,\ldots,v_k,w_1\circ\pi,\ldots,w_e\circ\pi)$ be a system of local co-ordinates for (V,v_0) (such a system exists because π is a submersion). Then, by (1.2)

$$m_V = \langle v_1,\ldots,v_k,w_1\circ\pi,\ldots,w_e\circ\pi\rangle.C_V$$

Now $i^* : C_V \to C_U$ acts as follows:

$$i^*(1) = 1 \text{ , } i^*(v_j) = v_j\circ i \text{ , } i^*(w_j\circ\pi) = w_j\circ\pi\circ i = 0 \text{ .}$$

Thus it is clear that $(v_1\circ i,\ldots,v_k\circ i)$ is a system of local co-ordinates for (U,u_0) , so, by (1.2), $v_1\circ i,\ldots,v_k\circ i$ form a basis for m_U .

Hence i^* is onto.

Also, clearly $\mathrm{Ker}\, i^* = \langle w_1\circ\pi,\ldots,w_e\circ\pi\rangle.C_V = \pi^*(m_W).C_V$, so that the sequence is indeed exact.

And of course the sequence is split via our choice of co-ordinates.

(1.5) Now observe that θ_N is a free C_N-module of rank = dim.N (if $(x_1,\ldots,x_n)$ is a system of local co-ordinates for (N,x_0) , the vector fields $\frac{\partial}{\partial x_1},\ldots,\frac{\partial}{\partial x_n}$ are a basis for it).

Now we define the C_N-module θ_f , for any map-germ $f : (N,x_0) \to (P,y_0)$, to be the module of germs of vector fields along f (that is, germs

$\phi : (N,x_0) \to TP$ such that the following diagram commutes:

$$\begin{array}{ccc} & & TP \\ & \overset{\phi}{\nearrow} & \Big\downarrow \text{projection} \\ (N,x_0) & \xrightarrow{f} & (P,y_0) \end{array} \qquad);$$

the C_N-module action is defined by $(\alpha.\phi)_x = \alpha_x.\phi_x$. It is clear that θ_f is a free C_N-module of rank $= \dim.P$ (if $(y_1,\ldots,y_p)$ is a system of local co-ordinates for (P,y_0), the vector fields $\frac{\partial}{\partial y_1}\circ f,\ldots,\frac{\partial}{\partial y_p}\circ f$ along f are a basis for it).

It is worth noting that θ_N is the same object as θ_{1_N}.

(1.6) There is a C_N-homomorphism $tf : \theta_N \to \theta_f$ defined by $tf(\xi) = Tf \circ \xi$; and there is also a homomorphism $wf : \theta_P \to \theta_f$ over f^* defined by $wf(\eta) = \eta \circ f$. (if A is an R-module, B an S-module, and $\phi : R \to S$ a ring homomorphism, then $\Phi : A \to B$ is a homomorphism <u>over</u> ϕ if $\Phi(\alpha a + \beta b) = \phi(\alpha).\Phi(a) + \phi(\beta).\Phi(b)$ for all $\alpha,\beta \in R$, $a,b \in A$).

(1.7) We may interpret θ_f as 'infinitesimal perturbations' of f : in some sense, elements of θ_f are 'tangent vectors' at f in the space of germs, specifying as they do for each $x \in N$ how $f(x)$ is to be deformed by giving a tangent vector there. We should thus expect to be able to derive elements of θ_f from paths in the germ-space, which we do as follows:-

if $g : (N \times U, x_0 \times u_0) \to (P,y_0)$ is such that $f(x) = g(x,u_0)$, then for each $\partial \in TU_{u_0}$ there is a well-defined element $\partial\cdot g \in \theta_f$ defined as follows: if $c : (\mathbb{R},0) \to (U,u_0)$ represents ∂, define $\partial.g(x) = \frac{d}{dt}(g(x,c(t)))\big|_{t=0}$. It is immediate (by Taylor's theorem and differentiation of products) that $\partial\cdot g$ is well-defined.

Such parametrized families of germs give rise to unfoldings $F : (N \times U, x_0 \times u_0) \to (P \times U, y_0 \times u_0)$ of f by $F(x,u) = (g(x,u),u)$, and vice versa; it is worth noting that

$$\partial.g = \{tF(0,\partial \circ p_N) - wF(0,\partial \circ p_P)\}|N \times \{u_0\}$$

(where we identify $\theta_{N\times U} = \theta p_N \oplus \theta p_u$, $\theta_{P\times U} = \theta p_P \oplus \theta p_U$).

Conversely, given an element of θ_f, can we construct a parametrized family of germs (and hence an unfolding) realising it? As to this, we have:

(1.8) Lemma

If $\phi_1,\ldots,\phi_k \in \theta_f$, then there exists a germ

$$g : (N \times \mathbb{R}^k, x_0 \times 0) \to (P, y_0)$$

such that $g(x,0) = f(x)$ and $\frac{\partial}{\partial u_i} \cdot g = \phi_i$ $(i = 1,\ldots,k)$.

Proof

Let $(y_1,\ldots,y_p)$ be a system of local co-ordinates for (P,y_0). Then we can write $\phi_i = \sum_{j=1}^{p} \phi i,j \cdot (\frac{\partial}{\partial y_j} \circ f)$ $(i=1,\ldots,k)$, where $\phi_{i,j} \in C_N$. Now define $g : (N \times \mathbb{R}^k, x_0 \times 0) \to (P,y_0)$ by

$$y_i \circ g(x,u_1,\ldots,u_k) = y_i \circ f(x) + \sum_{j=1}^{k} u_j \phi_{j,i}(x) .$$

Then clearly $g(x,0) = f(x)$, and $\frac{\partial}{\partial u_i} \cdot g = \phi_i$ $(i=1,\ldots,k)$, as required.

(1.9) Having introduced some of the modules with which we will be concerned, as well as indicating some of their relationship with unfoldings, we proceed to the theorems of module algebra which we will need.

(1.10) Lemma (Nakayama)

Let R be a commutative ring with identity, and let $m \subset R$ be an ideal such that $1 + m \subset R^*$ (R^* is the group of units of R).

Let E be a finitely-generated R-module.

Then if $m.E = E$, $E = 0$.

Proof

Suppose that $E \neq 0$, so that, since E is finitely-generated, there is a minimal spanning set $\{e_1,\ldots,e_k\}$ for E over R. Since $mE = E$, we can write $e_1 = \sum_{j=1}^{k} x_j \cdot e_j$, where $x_j \in m$. Thus $(1 - x_1).e_1 = \sum_{j=2}^{k} x_j e_j$, and so, since $1 - x_1$ is invertible, $e_1 = \sum_{j=2}^{k} (1 - x_1)^{-1} x_j . e_j$.

So e_1 is dependent on $e_2,\dots,e_k$ in contradiction to $\{e_1,\dots,e_k\}$ being minimal. Thus $E = 0$.

(1.11) Corollary

Let A be a finitely-generated R-module (R as in (1.10)), and let $S \subset A$ be a finite subset.

Then S generates A over R if and only if the image of S in A/mA spans A/mA over R/m.

Proof

'Only if': If S generates A over R, we have $R.S = A$.

Thus $(R/m).S \cong R.(S/mS) = A/mA$, so that indeed the image of S in A/mA spans A/mA over R/m.

'If': We have $(R/m).S = A/mA$, whence $A = m.A + R.S$, so that $A/R.S = m.(A/R.S)$. Thus, by (1.10) $A/R.S = 0$, or $A = R.S$. Thus S generates A over R.

Remark

Our use for (1.10), (1.11) is precisely that $R = C_N$, $m = m_N$ satisfy the conditions of (1.10) (the inverse of $1 + f$, for $f \in m_N$ is of course given by $(1 + f)^{-1}(x) = 1/(1 + f(x))$).

(1.12) Corollary

Let B be a sub-module of a finitely-generated C_N-module A, and suppose that

$$\dim_{\mathbb{R}} \{A/m_N^{r+1}A + B\} \leq r .$$

Then $m_N^r A \subset B$.

Proof

The non-decreasing sequence

$$\dim_{\mathbb{R}} \{A/A + B\} \leq \dots \leq \dim_{\mathbb{R}} \{A/m_N^i A + B\} \leq \dots \leq \dim_{\mathbb{R}} \{A/m_N^{r+1} A + B\}$$

consists of $r + 2$ non-negative integers, and is bounded above by r, by hypothesis. Hence there exists an integer i, $0 \leq i \leq r$, such that $\dim_{\mathbb{R}} \{A/m_N^i A + B\} = \dim_{\mathbb{R}} \{A/m_N^{i+1} A + B\}$, whence $m_N^i A + B = m_N^{i+1} A + B$.

Thus $(m_N^i A+B)/B = (m_N^{i+1}A+B)/B = m_N \cdot (m_N^i A+B)/B$, and so, by (1.10), $(m_N^i A+B)/B = 0$. Hence $m_N^i A \subset B$.

Now let us observe that if $f : (N,x_0) \to (P,y_0)$ is a map-germ and if A is a C_N-module, then A may also be considered as a C_P-module via f^* (that is, if $a \in A, \phi \in C_P$, we define $\phi . a = f^*(\phi) . a$). It will often be important for us, in this situation, to find generators for A as a C_P-module. As to this, we have, firstly:

(1.13) Theorem (Malgrange)

Let $f : (N,x_0) \to (P,y_0)$ be a map-germ, and let A be a finitely-generated C_N-module. Then A is finitely-generated as a C_P-module if and only if $\dim_{\mathbb{R}} \{A/f^*(m_P)A\}$ is finite.

This may look like a theorem of algebra; in fact, however, it is a very deep result in analysis, first proved by Malgrange (at the instigation of Thom). Another, slightly easier, proof is due to Mather [Ma IV]. We will not give a proof here.

(1.14) Corollary

Let A be a finitely-generated C_N-module, and let $S \subset A$ be a finite subset. Then S generates A as a C_P-module if and only if the image of S in $A/(f^*m_P)A$ spans $A/(f^*m_P)A$ over $\mathbb{R}\,(\equiv C_P/m_P)$.

Proof

'Only if' : clear.

'If' : If the image of S spans $A/(f^*m_P).A$ over $\mathbb{R}$, then $\dim_{\mathbb{R}}\{A/(f^*m_P)A\}$ is finite, so by (1.13) A is finitely-generated as a C_P-module. It then follows by (1.11) that S generates A over C_P .

(1.15) Corollary

Let E be a finitely-generated C_N-module, and let $E' \subset E$ be a finitely-generated sub-C_P-module (via f^*).

If

$$E = E' + (f^*m_P + m_N^{r+1})E$$

for some $r \geq \dim_{\mathbb{R}}\{E'/m_P E'\}$, then $E = E'$.

<u>Proof</u>

The inclusion $E' \subset E$ induces a vector-space homomorphism

$$E'/m_P E' \to E/(f^*m_P + m_N^{r+1})E ,$$

which is onto by our hypothesis, so that

$$\dim_{\mathbb{R}}\{E/(f^*m_P + m_N^{r+1})E\} \leq r .$$

So, since E is a finitely-generated C_N-module, it follows from (1.12) that $m_N^r E \subset (f^*m_P).E$.

Thus the hypothesis may now be written

$$E = E' + (f^*m_P)E .$$

So the vector-space homomorphism $E'/m_P E' \to E/(f^*m_P)E$ induced by inclusion is onto, and thus by (1.13) E is finitely-generated as a C_P-module.

But the reqritten hypothesis also implies $E/E' = (f^*m_P).(E/E')$. So, by (1.10), $E/E' = 0$, or $E = E'$.

§2. Infinitesimal Stability

In this section we state and prove an algebraic condition equivalent to stability, and derive various consequences. The condition is:

(2.1) Definition

A map-germ $f : (N,x_0) \to (P,y_0)$ is infinitesimally stable if

$$\theta_f = tf(\theta_N) + wf(\theta_P)$$

(for notation see (1.5), (1.6)).

The main result of the section is:

(2.2) Theorem

A map-germ is stable if and only if it is infinitesimally stable.

Proof

'Only if' : Suppose $f : (N,x_0) \to (P,y_0)$ is a stable map-germ.

Let $\phi \in \theta_f$, and let $g : (N \times \mathbb{R}, x_0 \times 0) \to (P,y_0)$ be such that $g(x,0) = f(x)$ and $\frac{\partial}{\partial u} \cdot g = \phi$ (as in (1.8)). Then g determines an unfolding $(F, 1_N \times 0, 1_P \times 0)$ of f by $f(x,u) = (g(x,u),u)$, which, since f is stable, must be trivial: let r,s be retractions to $1_N \times 0$, $1_P \times 0$ respectively such that $f \circ r = s \circ F$.

Let $\xi = \frac{\partial}{\partial u}\{r(x,u)\}|_{u=0} \in \theta_N$, $\eta = \frac{\partial}{\partial u}\{s(y,u)\}|_{u=0} \in \theta_P$. Differentiating the identity $f \circ r(x,u) = s(g(x,u),u)$ with respect to u, we obtain $tf(\xi) = \phi + wf(\eta)$, so that $\phi \in tf(\theta_N) + wf(\theta_P)$.

Since this is true for any $\phi \in \theta_f$, f is infinitesimally stable.

Before proving the 'If' part of the theorem, it will be convenient for us to derive a consequence of infinitesimal stability relating to vector fields along unfoldings.

Let $f : (N,x_0) \to (P,y_0)$ be a map-germ, (F,i,j) an unfolding of f, and $\pi : (P',y_0') \to (U,u_0)$ a submersion such that $\pi^{-1}(u_0) = \mathrm{Im}(j)$.

Let $\theta_{F,\pi} = \mathrm{Ker}\{d\pi : \theta_F \to \theta_{\pi \circ F}\}$

(where $d\pi$ is defined by $d\pi(\phi) = T\pi \circ \phi$).

Then there is a $C_{N'}$-module morphism $\alpha : \theta_{F,\pi} \to \theta_f$ (where θ_f is considered as a $C_{N'}$-module via i^*) defined by $\phi \to Tj^{-1} \circ \phi \circ i$ (this makes sense because $\phi(i(x)) \subset \mathrm{Ker}\ T\pi_{F(i(x))} = \mathrm{Im}(Tj_{f(x)})$).

Further,

$$0 \to (\pi \circ F)^* m_u \cdot \theta_{F,\pi} \subset \theta_{F,\pi} \xrightarrow{\alpha} \theta_f \to 0$$

is an exact sequence of $C_{N'}$-modules.

(If $\phi \in \theta_f$, then $dj \circ \phi$ is a vector field along $j \circ f = F \circ i$, so that $Tj \circ \phi \circ i^{-1} : (\mathrm{Im}(i), x_0') \to TP'$ is a vector field along $F|\mathrm{Im}(i)$. Since $\mathrm{Im}(i)$ is a submanifold of N' and $\mathrm{Im}(\phi) \subset \mathrm{Im}\ Tj = \mathrm{Ker}\ T\pi$, this may be extended to a germ ϕ' of vector field along F with $\mathrm{Im}(\phi') \subset \mathrm{Ker}\ T\pi$; and then $\alpha(\phi') = \phi$. Thus α is onto.

Also,

$$\begin{aligned}\mathrm{Ker}\ \alpha &= \{\phi \in \theta_{F,\pi} \mid dj^{-1} \circ \phi \circ i = 0\} \\ &= \{\phi \in \theta_{F,\pi} \mid \phi \circ i = 0\} \\ &= (\mathrm{Ker}\ i^*)\theta_{F,\pi}\end{aligned}$$

But, by (1.4), $(\mathrm{Ker}\ i^*)C_{N'} = (\pi \circ F)^* m_u \cdot C_{N'}$; so the sequence is indeed exact).

Thus there is an isomorphism of $C_{N'}$-modules

(a) $$\theta_f \cong \theta_{F,\pi} / (\pi \circ F)^* m_u \cdot \theta_{F,\pi}$$

Similarly (replacing f by $1_N, 1_P$ respectively), we have isomorphisms

(b) $\theta_N \cong \theta_{N',\pi\circ F} / (\pi \circ F)^* m_u \cdot \theta_{N',\pi\circ F}$ (where $\theta_{N',\pi\circ F} = \mathrm{Ker}\ t(\pi \circ F)$)

(c) $\theta_P \cong \theta_{P',\pi} / (\pi^* m_u) \cdot \theta_{P',\pi}$ (where $\theta_{P',\pi} = \mathrm{Ker}\ t\pi$).

(2.3) <u>Lemma</u>

If f is infinitesimally stable, then

$$\theta_{F,\pi} = tF(\theta_{N',\pi\circ F}) + wF(\theta_{P',\pi})$$

<u>Proof</u>

It is easy to see that the morphism $\alpha : \theta_{F,\pi} \to \theta_f$ defined above maps $tF(\theta_{N',\pi\circ F})$ onto $tf(\theta_N)$ and $(F^* m_{P'})\theta_{F,\pi}$ onto $(f^* m_P)\theta_f$. Hence, since $(\pi \circ F)^* m_u = F^*(\pi^* m_u) \subset F^* m_{P'}$, we can deduce from (a) an isomorphism of $C_{N'}$-modules

(*) $$\theta_{F,\pi} / tF(\theta_{N',\pi\circ F}) + (F^* m_{P'})\theta_{F,\pi} \cong \theta_f / tf(\theta_N) + (f^* m_P)\theta_f$$

Since f is infinitesimally stable, $\theta_f = tf(\theta_N) + wf(\theta_P)$, so that a finite $\mathbb{R}$-basis for the RHS of (*) is given by the projection of $wf(S)$, for some finite $S \subset \theta_P$.

By (c), S is the projection of a finite set $S' \subset \theta_{P',\pi}$; and then, by the isomorphism (*) , $wF(S')$ must project to an $\mathbb{R}$-basis for the LHS of (*) . Thus, by (1.14) applied to the $C_{N'}$-module $\theta_{F,\pi}/tF(\theta_{N',\pi\circ F})$, the projection of $wF(S')$ must generate this $C_{N'}$-module as a $C_{P'}$-module. Certainly, then, $\theta_{F,\pi} = tF(\theta_{N',\pi\circ F}) + wF(\theta_{P',\pi})$.

Now we can give:

<u>Proof</u> of 'If' in (2.2)

Let $f : (N,x_0) \to (P,y_0)$ be infinitesimally stable, and let (F,i,j) be any unfolding of f . To show that f is stable, we must show that (F,i,j) is a trivial unfolding; by (0.6) it will suffice to construct, for any $\tau \in \theta_U$, vector fields $\xi \in \theta_{N'}$, $\eta \in \theta_{P'}$ such that $tF(\xi) = wF(\eta)$ and $t\pi(\eta) = w\pi(\tau)$ (where $\pi : (P',y_0') \to (U,u_0)$ is a submersion such that $\pi^{-1}(u_0) = \mathrm{Im}(j)$, as usual).

So - let $\tau \in \theta_u$, and let $\xi_1 \in \theta_{N'}, \eta_1 \in \theta_{P'}$ be lifts for τ over $\pi\circ F$ and π respectively, that is,

$$T(\pi\circ F)\circ\xi_1 = \tau\circ(\pi\circ F)\ , \ T\pi\circ\eta_1 = \tau\circ\pi$$

(Such lifts exist because $\pi\circ F, \pi$ are submersions).

Then $T\pi\circ\eta_1\circ F - T\pi\circ dF\circ\xi_1 = 0$, so

$$wF(\eta_1) - tF(\xi_1) \in \mathrm{Ker}\ d\pi = \theta_{F,\pi}$$

Thus, by (2.3), there exist $\xi_2 \in \theta_{N',\pi\circ F}$, $\eta_2 \in \theta_{P',\pi}$ such that

$$wF(\eta_1) - tF(\xi_1) = wF(\eta_2) + tF(\xi_2)\ .$$

Define $\xi = \xi_1 + \xi_2$, $\eta = \eta_1 - \eta_2$.

Then $tF(\xi) = tF(\xi_1) + tF(\xi_2) = wF(\eta_1) - wF(\eta_2) = wF(\eta)$ and $t\pi(\eta) = t\pi(\eta_1) - t\pi(\eta_2) = \tau\circ\pi = w\pi(\tau)$.

Thus ξ,η are the required vector fields, and the proof is complete.

(2.4)' Let $f : (N,x_0) \to (P,y_0)$ be a map-germ, and let $\mathcal{N}_f$ be the $\mathbb{R}$-vector space $\theta_f/tf(\theta_N) + (f^*m_P)\theta_f$.

Let $\rho_f : \theta_P/m_P\theta_P \to \mathcal{N}_f$ be the $\mathbb{R}$-homomorphism induced by wf (this is well-defined since $wf(m_P\theta_P) \subset (f^*m_P)\theta_f$).

Then:

Corollary

f is stable if and only if ρ_f is surjective.

Proof

'Only if' : clear.

'If' : If ρ_f is surjective, then

$$\theta_f = tf(\theta_N) + wf(\theta_P) + (f^*m_P)\theta_f .$$

Then by (1.15) (with $E = \theta_f$, $E' = tf(\theta_N) + wf(\theta_P)$) we have $\theta_f = tf(\theta_N) + wf(\theta_P)$, so that f is infinitesimally stable, and hence, by (2.2), stable.

We want to go on from this to derive a condition under which a map-germ f will possess a stable unfolding (F,i,j). Of course F is stable if and only if ρ_F is surjective; we want to express this in terms of the module θ_f. As a first step, we have

(2.5) Lemma

There is an $\mathbb{R}$-linear isomorphism

$$q_{F,f} : \mathcal{N}_f \to \mathcal{N}_F$$

defined as follows:

if $\phi \in \theta_f$, let $\Phi \in \theta_F$ be such that $\Phi \circ i = dj \circ \phi$.

Then
$$q_{F,f}([\phi]) = [\Phi] .$$

Proof

Let $\pi : (P',y_0') \to (U,u_0)$ be a submersion such that $\pi^{-1}(u_0) = \mathrm{Im}(j)$. Consider the $C_{N'}$-module morphism

$$\theta_{F,\pi}/tF(\theta_{N',\pi\circ F}) \to \theta_F/tF(\theta_{N'})$$

induced by the inclusion $\theta_{F,\pi} \subset \theta_F$. This morphism is injective, because

$$\theta_{F,\pi} \cap tF(\theta_{N'}) = \text{Ker } d\pi \cap \text{Im } tF = tF(\ker t(\pi \circ F)) = tF(\theta_{N',\pi\circ F}) .$$

On the other hand, we have $\theta_F = \theta_{F,\pi} + tF(\theta_{N'})$ (for if $(u_1,\ldots,u_k)$ is a system of local co-ordinates at (U,u_0), then we may choose a system of local co-ordinates at (P',y_0') of form $(y_1,\ldots,y_p,u_1 \circ \pi,\ldots,u_k \circ \pi)$ such that $\{\frac{\partial}{\partial y_i}\}$ spans Ker $t\pi$ (so that $\{wF(\frac{\partial}{\partial y_i})\}$ spans $\theta_{F,\pi}$). Then $\{wF(\frac{\partial}{\partial y_i}), wF(\frac{\partial}{\partial (u_j \circ \pi)})\}$ spans θ_F. But

$$d\pi(wF(\frac{\partial}{\partial(u_j\circ\pi)}) - tF(\frac{\partial}{\partial(u_j\circ\pi\circ F)}) = 0$$

so that also $\{wF(\frac{\partial}{\partial y_i}), tF(\frac{\partial}{\partial(u_j\circ\pi\circ F)})\}$ spans θ_F).

Thus the $C_{N'}$-morphism is actually an isomorphism. It is, therefore, also an isomorphism of $C_{P'}$-modules (via F^*), and so gives rise to an $\mathbb{R}$-linear isomorphism

$$\theta_{F,\pi}/tF(\theta_{N',\pi\circ F}) + (F^*m_{P'})\theta_{F,\pi} \cong \theta_F/tF(\theta_{N'}) + (F^*m_{P'})\theta_F .$$

Composing the inverse of this isomorphism with the isomorphism (*) derived in (2.3), we obtain $q_{F,f}$.

The characterization of the action of $q_{F,f}$ follows from the definition of the isomorphism (*) of (2.3), this isomorphism being induced by the morphism $\alpha : \theta_{F,\pi} \to \theta_f$ defined by $\alpha(\phi) = dj^{-1} \circ \phi \circ i$.

(2.6) We define the $\mathbb{R}$-homomorphism

$$\rho_{F,f} : TP'_{y_0'} \to \mathcal{N}_f$$

as the composite

$$TP'_{y_0'} \equiv \theta_{P'}/m_{P'}\theta_{P'} \xrightarrow{\rho_F} \mathcal{N}_F \xrightarrow{q^{-1}_{F,f}} \mathcal{N}_f .$$

We have

<u>Lemma</u>

(F,i,j) is a stable unfolding of f if and only if $\rho_{F,f}$ is surjective.

<u>Proof</u>

Immediate from (2.4) and (2.5).

This leads us to

(2.7) Definition

Let $f : (N,x_0) \to (P,y_0)$ be a map-germ.

Define $\chi_f = \dim_{\mathbb{R}} \mathcal{N}_f \in \{0,1,2,\ldots,\infty\}$.

We say that f is of finite singularity type (abbreviated FST) if $\chi_f < \infty$.

(2.8) Theorem

f has a stable unfolding if and only if f is of FST.

Proof

'Only if' : If (F,i,j) is a stable unfolding of f , then $\rho_{F,f}$ is surjective, so that certainly $\chi_f \le \dim P'$.

'If' : If $k = \chi_f < \infty$, let $g : (N \times \mathbb{R}^k, x_0 \times 0) \to (P,y_0)$ be such that $g(x,0) = f(x)$, and $\{ \frac{\partial}{\partial u_i} \cdot g\}_{i=1}^k$ maps to an $\mathbb{R}$-basis in $\mathcal{N}_f$ (such g exists by (1.8)). Then g determines an unfolding $(F,1_N \times 0,1_P \times 0)$ of f , defined by $F(x,u) = (g(x,u),u)$.

We have $\rho_{F,f}(\frac{\partial}{\partial u_i}|y_0) = q_{F,f}^{-1}[wF(\frac{\partial}{\partial u_i})]$.

Also, $\frac{\partial}{\partial u_i} \cdot g = \{tF(\frac{\partial}{\partial u_i}) - wF(\frac{\partial}{\partial u_i})\}|N \times 0$ (see (1.7)), and so

$$[\frac{\partial}{\partial u_i} \cdot g] = -\rho_{F,f}(\frac{\partial}{\partial u_i}|y_0) .$$

So, since $\{[\frac{\partial}{\partial u_i} \cdot g]\}$ spans $\mathcal{N}_f$, it follows that $\rho_{F,f}$ is surjective, so that $(F,1_N \times 0,1_P \times 0)$ is a stable unfolding of f .

§3. Versality

In this section we show the equivalence of versality and stability for unfoldings.

In fact, the key result of this section is a uniqueness theorem:

(3.1) Theorem

Any two stable unfoldings of the same dimension of a map-germ f are isomorphic as unfoldings.

The main step in proving this is establishing a 'continuous' version of it, which is the following:

(3.2) Lemma

Suppose that the commutative diagram

$$\begin{array}{ccccc}
(N'\times\mathbb{R}, x_0'\times I) & & \xrightarrow{F} & & (P'\times\mathbb{R}, y_0'\times I) \\
& \searrow & & \swarrow{\scriptstyle \pi_{\mathbb{R}}} & \\
{\scriptstyle i\times 1_{\mathbb{R}}}\uparrow & & (\mathbb{R}, I) & & \uparrow{\scriptstyle j\times 1_{\mathbb{R}}} \\
& \nearrow & & \nwarrow & \\
(N\times\mathbb{R}, x_0\times I) & & \xrightarrow{f\times 1_{\mathbb{R}}} & & (P\times\mathbb{R}, y_0\times I)
\end{array}$$

is a one-parameter family of stable unfoldings of a map-germ f (that is, for each $a \in I$ (I is the unit interval $[0,1]$) the unfolding (F^a,i,j) of f, where F^a is defined by $F(x',a) = (F^a(x'),a)$, is stable).

Then the family is trivial over $(\mathbb{R},I)$ (that is, there exist retractions r,s to $1_{N'}\times 0, 1_{P'}\times 0$ respectively such that the whole diagram

$$\begin{array}{ccccc}
(N', x_0') & & \xrightarrow{F^0} & & (P', y_0') \\
\uparrow{\scriptstyle r} & & & & \uparrow{\scriptstyle s} \\
(N'\times\mathbb{R}, x_0'\times I) & & \xrightarrow{F} & & (P'\times\mathbb{R}, y_0'\times I) \\
& \searrow & & \swarrow{\scriptstyle \pi_{\mathbb{R}}} & \\
{\scriptstyle i\times 1_{\mathbb{R}}}\uparrow & & (\mathbb{R}, I) & & \uparrow{\scriptstyle j\times 1_{\mathbb{R}}} \\
& \nearrow & & \nwarrow & \\
(N\times\mathbb{R}, x_0\times I) & & \xrightarrow{f\times 1_{\mathbb{R}}} & & (P\times\mathbb{R}, y_0\times I)
\end{array} \qquad (*)$$

commutes).

Proof

We will make the natural identifications

$$\theta_{N'\times\mathbb{R},x_0'\times a} = \theta_{p_{N'}} \oplus \theta_{\pi_{\mathbb{R}}\circ F}$$

$$\theta_{P'\times\mathbb{R},y_0'\times a} = \theta_{p_{P'}} \oplus \theta_{\pi_{\mathbb{R}}}$$

(where, as usual $p_{N'} : (N' \times \mathbb{R}, x_0' \times a) \to (N', x_0')$ is the natural projection; and similarly for $p_{P'}$).

We claim that to construct r,s it will be sufficient to find, for each $a \in I$,

$$\xi_a = (\xi_a', \frac{\partial}{\partial u}\circ \pi_{\mathbb{R}}\circ F) \in \theta_{N'\times\mathbb{R},x_0'\times a}$$

$$\eta_a = (\eta_a', \frac{\partial}{\partial u}\circ \pi_{\mathbb{R}}) \in \theta_{P'\times\mathbb{R},y_0'\times a}$$

such that ξ_a is a lift for η_a over F_a (where $F_a : (N' \times \mathbb{R}, x_0' \times a) \to (P' \times \mathbb{R}, y_0' \times a)$ is the germ-restriction of F) , where

$$\xi_a' \in \mathrm{Ker}(i \times 1_{\mathbb{R}})^*\theta_{p_{N'}}$$

$$\eta_a' \in \mathrm{Ker}(j \times 1_{\mathbb{R}})^*\theta_{p_{P'}}$$

(η_a is, of course, automatically a lift for $\frac{\partial}{\partial u}$ over $\pi_{\mathbb{R}}$)

To see this, we proceed as follows:-

let $\qquad \phi_a : ((N' \times \mathbb{R}) \times \mathbb{R}, (x_0' \times a) \times 0) \to (N' \times \mathbb{R}, x_0' \times a)$

$\qquad\quad \psi_a : ((P' \times \mathbb{R}) \times \mathbb{R}, (y_0' \times a) \times 0) \to (P' \times \mathbb{R}, y_0' \times a)$

be flows for ξ_a, η_a respectively. We may in fact suppose (via choice of representatives for F, ξ_a, η_a) that ϕ_a, ψ_a are germs defined on neighbourhoods of $x_0' \times [a, a+\varepsilon] \times [-\varepsilon, \varepsilon]$, $y_0' \times [a, a+\varepsilon] \times [-\varepsilon, \varepsilon]$ respectively, for some $\varepsilon > 0$. Hence, since I is compact and connected, there exists an increasing sequence $0 = a_0 < \ldots < a_e = 1$ such that ϕ_{a_i}, ψ_{a_i} are germs defined on neighbourhoods of $x_0' \times [a_i, a_{i+1}] \times [a_i - a_{i+1}, a_{i+1} - a_i]$, $y_0' \times [a_i, a_{i+1}] \times [a_i - a_{i+1}, a_{i+1} - a_i]$ respectively.

Then r,s may be defined by

$$(r(x',a_i+t),0) = \phi_{a_0,a_0-a_1} \circ \phi_{a_1,a_1-a_2} \circ \ldots \circ \phi_{a_{i-1},a_{i-1}-a_i} \circ \phi_{a_i,a_i-t}(x',a_i+t)$$

$$(s(y',a_i+t),0) = \psi_{a_0,a_0-a_1} \circ \psi_{a_0,a_0-a_1} \circ \ldots \circ \psi_{a_{i-1},a_{i-1}-a_i} \circ \psi_{a_i,a_{i-t}}(y',a_i+t)$$

for $0 \le t \le a_{i+1}-a_i$, and (with i replaced by $0,e$ respectively) for t near $a_0 = 0, a_e = 1$.

Of course $\mathrm{Ker}(i \times 1_{\mathbb{R}})^* \subset p_{N'}^* m_{N'}$, $\mathrm{Ker}(j \times 1_{\mathbb{R}})^* \subset p_{P'}^* m_{P'}$, so that $\xi_a' \in (p_{N'}^* m_{N'}) \cdot \theta_{p_{N'}}$, $\eta_a' \in (p_{P'}^* m_{P'}) \cdot \theta_{p_{P'}}$. Thus we will have $\phi_{a,t}(x_0',a+t') = (x_0',a+t'+t)$, $\psi_{a,t}(y_0',a+t') = (y_0',a+t'+t)$ so that $r(x_0' \times I) = x_0'$, $s(y_0' \times I) = y_0'$. Then arguments similar to those of (0.6) show that the top pentagon of (*) commutes.

More strongly, the facts that $\xi_a' \in \mathrm{Ker}\,(i \times 1_{\mathbb{R}})^* \theta_{p_{N'}}$, $\eta_a' \in \mathrm{Ker}(j \times 1_{\mathbb{R}})^* \theta_{p_{P'}}$ will ensure that $r|i(N) \times \mathbb{R}$, $s|j(P) \times \mathbb{R}$ are the natural projections to $i(N), j(P)$ respectively, and hence that the whole of (*) commutes.

Thus it remains to construct ξ_a, η_a for each a.

Let

$$\phi_a = tF_a(0, \frac{\partial}{\partial u} \circ \pi_{\mathbb{R}} \circ F_a) - wF_a(0, \frac{\partial}{\partial u} \circ \pi_{\mathbb{R}}) .$$

Then it is easy to see that $T\pi_{\mathbb{R}} \circ \phi_a = 0$, and that $\phi_a \circ (i \times 1_{\mathbb{R}}) = 0$, so that $\phi_a \in \mathrm{Ker}(i \times 1_{\mathbb{R}})^* \theta_{F_a, \pi_{\mathbb{R}}}$.

Now, since F^a is stable for any $a \in I$, and since $(F_a, 1_{N'} \times 0, 1_{P'} \times 0)$ is an unfolding of F^a, we have by (2.3) that

$$\theta_{F_a,\pi_{\mathbb{R}}} = tF_a(\theta_{N' \times \mathbb{R}, \pi_{\mathbb{R}} \circ F_a}) + wF_a(\theta_{P' \times \mathbb{R}, \pi_{\mathbb{R}}})$$

and thus

$$\mathrm{Ker}(i \times 1_{\mathbb{R}})^* \theta_{F_a,\pi_{\mathbb{R}}} \subset tF\,(\mathrm{Ker}(i \times 1_{\mathbb{R}})^* \theta_{N' \times \mathbb{R}, \pi_{\mathbb{R}} \circ F_a}) + wF_a(\mathrm{Ker}(j \times 1_{\mathbb{R}})^* \theta_{P' \times \mathbb{R}, \pi_{\mathbb{R}}})$$

Hence there exist $\xi_a' \in \mathrm{Ker}(i \times 1_{\mathbb{R}})^* \theta_{N' \times \mathbb{R}, \pi_{\mathbb{R}} \circ F_a}$, $\eta_a' \in \mathrm{Ker}(j \times 1_{\mathbb{R}})^* \theta_{P' \times \mathbb{R}, \pi_{\mathbb{R}}}$ such that $\phi_a = -tF_a(\xi_a') + wF_a(\eta_a')$.

Now observe that the identification $\theta_{N' \times \mathbb{R}} = \theta p_{N'} \oplus \theta_{\pi_{\mathbb{R}} \circ F_a}$ identifies $\theta p_{N'}$ with $\theta_{N' \times \mathbb{R}, \pi_{\mathbb{R}} \circ F_a}$, and similarly $\theta p_{P'}$ is identified with $\theta_{P' \times \mathbb{R}, \pi_{\mathbb{R}}}$.

With respect to these identifications, define

$$\xi_a = (\xi'_a , \frac{\partial}{\partial u}\circ \pi_{\mathbb{R}}\circ F_a) \ , \ \eta_a = (\eta'_a , \frac{\partial}{\partial u}\circ \pi_{\mathbb{R}}) \ .$$

Then

$$tF_a(\xi_a) - wF_a(\eta_a) = \phi_a - (-tF_a(\xi'_a) + wF_a(\eta'_a)) = 0$$

so that ξ_a is a lift for η_a over F_a.

Thus ξ_a, η_a have the required form, and the proof is complete.

Now we can give

<u>Proof of (3.1)</u>

Let (F,i,j), (F',i',j') be two stable unfoldings of f of the same dimension k. By finding submersions $\pi : (P',y'_0) \to (U,0)$ such that $\pi^{-1}(0) = \mathrm{Im}(j)$, $\pi' : (P'',y''_0) \to (U,0)$ such that $\pi'^{-1}(0) = \mathrm{Im}(j')$, where U is a neighbourhood of 0 in $\mathbb{R}^k$, we may identify F, F' as level-preserving map-germs $(N\times U, x_0\times 0) \to (P\times U, y_0\times 0)$ such that $F(x,0) = (f(x),0) = F'(x,0)$, so that it remains to show that $(F, 1_N\times 0, 1_P\times 0)$, $(F', 1_N\times 0, 1_P\times 0)$ are isomorphic unfoldings.

Since F, F' are stable, the $\mathbb{R}$-homomorphisms

$$\rho_{F,f}, \rho_{F',f} : T(P\times U)_{y_0\times 0} \to \theta_f/tf(\theta_N) + (f^*m_P)\theta_f$$

are both surjective, and hence so also are the $\mathbb{R}$-homomorphisms

$$p_{F,f}, p_{F',f} : TU_0 \to \theta_f/tf(\theta_N) + wf(\theta_P) + (f^*m_P)\theta_f$$

induced by restriction and projection (recall that $\rho_{F,f}$ is induced by wF).

Thus there exists an isomorphism h of $TU_0 = \mathbb{R}^k$ such that $p_{F,f} = p_{F',f}\circ h$.

Let $g' : (N\times U, x_0\times 0) \to (P,y_0)$ be defined by $(x,g'(x,u)) = F'(x,u)$; then define $F'' : (N\times U, x_0\times 0) \to (P\times U, y_0\times 0)$ by $F''(x,u) = (g'(x,h(u)),u)$. F'' is an unfolding of f, and $(1_N\times h, 1_P\times h^{-1}) : (F', 1_N\times 0, 1_P\times 0) \to (F'', 1_N\times 0, 1_P\times 0)$ is an isomorphism of unfoldings, so that it remains to show that $(F, 1_N\times 0, 1_P\times 0)$, $(F'', 1_N\times 0, 1_P\times 0)$ are isomorphic.

We claim $p_{F'',f} = p_{F,f}$. To see this, recall that if $g : (N\times U, x_0\times 0) \to (P,y_0)$ is defined by $F(x,u) = (g(x,u),u)$, then (as in (2.8)) $p_{F,f}(\partial_0) = [-\partial\cdot g]$ for $\partial \in \theta_U$. Then (defining g'' from F'' in the same way as g, g' were defined from F, F') we have

$$p_{F'',f}(\partial_0) = [-\partial \cdot g''] = [-h(\partial) \cdot g'] = p_{F',f}(h(\partial_0)) = p_{F,f}(\partial)$$

and the claim is proved.

Now let $y_1, \ldots, y_p$ be a system of local co-ordinates for (P, y_0) , and define $g^a : (N \times U, x_0 \times 0) \to (P, y_0)$ $(a \in I)$ by

$$y_i \circ g^a(x,u) = (1-a) \cdot y_i \circ g(x,u) + a \cdot y_i \circ g''(x,u) \qquad (i = 1, \ldots, p) .$$

Then, defining $F^a : (N \times U, x_0 \times 0) \to (P \times U, y_0 \times 0)$ by $F^a(x,u) = (g^a(x,u), u)$, we clearly have $p_{F^a,f} = (1-a)p_{F,f} + ap_{F'',f} = p_{F,f}$. So in particular $p_{F^a,f}$ is surjective, and hence so also is $\rho_{F^a,f}$. Thus F^a is stable for each $a \in I$.

Now define the map-germ $\tilde{F} : (N \times U \times \mathbb{R}, x_0 \times 0 \times I) \to (P \times U \times \mathbb{R}, y_0 \times 0 \times I)$ by $\tilde{F}(x,u,a) = (F^a(x,u), a)$. Then, by (3.2), there exist retractions r, s to $1_{N \times U} \times 0, 1_{P \times U} \times 0$ respectively such that the following diagram commutes :

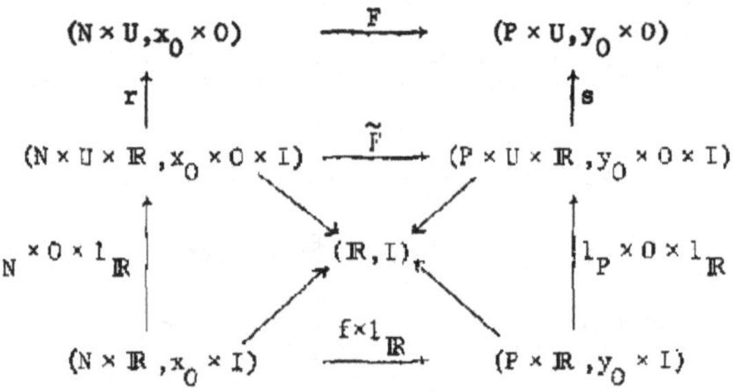

Define $r^1 : (N \times U, x_0 \times 0) \to (N \times U, x_0 \times 0)$ by $r^1(x,u) = r(x,u,1)$, and define $s^1 : (P \times U, y_0 \times 0) \to (P \times U, y_0 \times 0)$ by $s^1(y,u) = s(y,u,1)$. Then clearly $(r^1, s^1) : (F'', 1_N \times 0, 1_P \times 0) \to (F, 1_N \times 0, 1_P \times 0)$ is an isomorphism of unfoldings, and so the proof is complete.

(3.3) <u>Corollary</u>

Let $(F,i,j), (F',i',j')$ be stable unfoldings of f such that

$\dim.(F',i',j') - \dim.(F,i,j) = e \geq 0$.

Then (F',i',j') and $(F \times 1_{\mathbb{R}^e}, i \times 0, j \times 0)$ are isomorphic unfoldings.

<u>Proof</u>

By (3.1), we need only show that $(F \times 1_{\mathbb{R}^e}, i \times 0, j \times 0)$ is a stable unfolding of f .

Let $g : (N' \times \mathbb{R}^e, x_0' \times 0) \to (P', y_0')$ be defined by $(g(x',t),t) = (F \times 1_{\mathbb{R}^e})(x',t)$.

Then $\rho_{F\times 1_{\mathbb{R}^e},F}(\frac{\partial}{\partial t_i}\Big|_{y_0\times 0}) = [-\frac{\partial}{\partial t_i}\cdot g] = 0$ (where $t_1,\ldots,t_e$ are the usual co-ordinates in $\mathbb{R}^e$) so that $\rho_{F\times 1_{\mathbb{R}^e},F} = \rho_F \circ T\pi_{y_0'\times 0}$, where $\pi : (P' \times \mathbb{R}^e, y_0' \times 0) \to (P', y_0')$ is the natural projection.

Thus

$$\rho_{F\times 1_{\mathbb{R}^e},f} = (q_{F,f})^{-1} \circ \rho_{F\times 1_{\mathbb{R}^e},F} = \rho_{F,f} \circ T\pi_{y_0\times 0} .$$

Hence, since $\rho_{F,f}$, $T\pi_{y_0\times 0}$ are surjective, so is $\rho_{F\times 1_{\mathbb{R}^e},f}$; and thus $(F \times 1_{\mathbb{R}^e}, i \times 0, j \times 0)$ is indeed a stable unfolding of f .

Now we can prove

(3.4) <u>Theorem</u>

An unfolding of f is versal if and only if it is stable.

<u>Proof</u>

'If' : Suppose that (F,i,j) is a stable unfolding of f ; let (F',i',j') be any other unfolding of f . Since f has a stable unfolding, $\chi_f < \infty$. So, since $\chi_{F'} = \chi_f$ (via $q_{F',f}$), $\chi_{F'} < \infty$ and so F' has a stable unfolding, say (F'',I,J) . Then $(F'',I\circ i',J\circ j')$ is a stable unfolding of f .

Then, if $e = |\dim.(F'',I\circ i',J\circ j') - \dim (F,i,j)|$, we have, by (3.3), either an isomorphism

$$(\phi,\psi) : (F \times 1_{\mathbb{R}^e}, i \times 0, j \times 0) \approx (F'', I\circ i', J\circ j')$$

or an isomorphism

$$(\phi,\psi) : (F,i,j) \approx (F'' \times 1_{\mathbb{R}^e}, I\circ i' \times 0, J\circ j' \times 0) .$$

In the first case, $((\pi\circ F)\circ\phi\circ I, \pi\circ\psi\circ J)$ is a morphism $(F',i',j') \to (F,i,j)$ (where $\pi : (P' \times \mathbb{R}^e, y_0' \times 0) \to (P', y_0')$ is the natural projection); while in the second case, $(\phi\circ(1_{N''}\times 0)\circ I , \psi\circ(1_{P''}\times 0)\circ J)$ is a morphism $(F'.i',j') \to (F,i,j)$.

So (F,i,j) is versal.

'Only if' : Suppose (F,i,j) is a versal unfolding of f . The neatest way to see that (F,i,j) is then a stable unfolding is via the naturality of $\rho_{F,f}$, which we make explicit as follows :

(3.5) Sublemma

Let

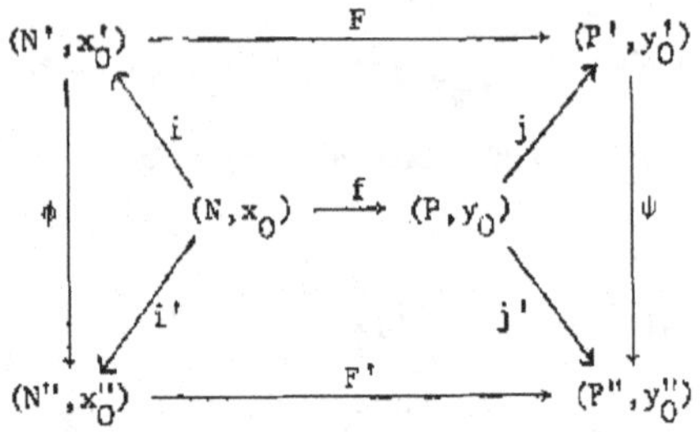

be a morphism $(\phi,\psi) : (F,i,j) \to (F',i',j')$ of unfoldings of f .

Then $\rho_{F,f} = \rho_{F',f} \circ T\psi_{y_0}$

Proof

Define an $\mathbb{R}$-homomorphism $K : \mathcal{N}_F \to \mathcal{N}_{F'}$ as follows : if $A \in \theta_F$, let $A' \in \theta_{F'}$ be such that $A' \circ i' = T\psi \circ A \circ i$. Then define $K([A]) = [A']$. This well-defined, for it is not hard to see that $q^{-1}_{F',f} \circ K = q^{-1}_{F,f}$. (for if $a \in \theta_f$, let $A \in \theta_F$ be such that $A \circ i = Tj \circ a$, and let $A' \in \theta_{F'}$ be such that $A' \circ i' = Tj' \circ a$. Then $q_{F,f}[a] = [A]$, $q_{F',f}[a] = [A']$. But $A' \circ i' = Tj' \circ a = T\psi \circ Tj \circ a = T\psi \circ A \circ i$, so $K([A]) = [A']$).

Now let $V \in \theta_{P'}$, and let $V' \in \theta_{P''}$ be such that $V' \circ j = T\psi \circ V \circ j$. Then

$$(V' \circ F') \circ i' = V' \circ \psi \circ F \circ i = V' \circ j' \circ f = T\psi \circ V \circ j \circ f = T\psi \circ (V \circ F) \circ i ,$$

so that $K([V \circ F]) = [V' \circ F']$, and thus

$$\rho_{F,f}(V_{y_0'}) = q^{-1}_{F,f}([V \circ F]) = q^{-1}_{F',f}([V' \circ F']) = \rho_{F',f}(V'_{y_0''}) = \rho_{F',f}(T\psi_{y_0'}(V_{y_0'})) .$$

Now we return to the proof of 'Only if' in (3.4).

Suppose that (F,i,j) is a versal unfolding of f , and suppose that F is not stable, so that $\rho_{F,f}$ is not surjective. Then let $\alpha \in \theta_f$ be such that $[\alpha] \notin \mathrm{Im}(\rho_{F,f})$, and let $g' : (N \times \mathbb{R}, x_0 \times 0) \to (P,y_0)$ be such that $g'(x,0) = f(x)$ and $\frac{\partial}{\partial u} \cdot g' = \alpha$ (such g' exists by (1.8)). Then define an unfolding $(F', 1_N \times 0, 1_P \times 0)$ of f by $F'(x,u) = (g'(x,u),u)$. Since (F,i,j) is versal, there exists a morphism $(\phi,\psi) : (F', 1_N \times 0, 1_P \times 0) \to (F,i,j)$, and, by (3.5), we have

$\rho_{F',f} = \rho_{F,f} \circ T\psi_{y_0 \times 0}$.

But $\rho_{F',f}(\frac{\partial}{\partial u}|_{y_0 \times 0}) = [-\frac{\partial}{\partial u} \cdot g] = [-\alpha]$, which is not contained in $\mathrm{Im}(\rho_{F,f})$

Thus we have arrived at a contradiction; so F must be stable.

§4. Contact-equivalence

Definition (4.1)

Map-germs $f_i : (N_i, x_i) \to (P_i, y_i)$ $(i = 0,1)$ are <u>equivalent</u> if there exist germs of diffeomorphism $h : (N_0, x_0) \to (N_1, x_1)$, $k : (P_0, y_0) \to (P_1, y_1)$ such that $f_1 \circ h = k \circ f_0$.

Definition (4.2)

a) A pair (h,H) consisting of germs of diffeomorphism

$$h : (N_0, x_0) \to (N_1, x_1),\ H : (N_0 \times P_0, x_0 \times y_0) \to (N_1 \times P_1, x_1 \times y_1)$$

such that the following diagram commutes:

$$\begin{array}{ccccc}
(N_0, x_0) & \xrightarrow{1_{N_0} \times y_0} & (N_0 \times P_0, x_0 \times y_0) & \xrightarrow{p_{N_0}} & (N_0, x_0) \\
\downarrow h & & \downarrow H & & \downarrow h \\
(N_1, x_1) & \xrightarrow{1_{N_1} \times y_1} & (N_1 \times P_1, x_1 \times y_1) & \xrightarrow{p_{N_1}} & (N_1, x_1)
\end{array}$$

is a <u>contact equivalence</u>.

If $N_0 = N_1, P_0 = P_1$, (h,H) is a <u>$\mathcal{K}$-equivalence</u>; if also $h = 1_{N_0}$, (h,H) is a <u>$\mathcal{C}$-equivalence</u>.

b) Map-germs $f_i : (N_i, x_i) \to (P_i, y_i)$ $(i = 0,1)$ are <u>contact-equivalent</u> (respectively <u>$\mathcal{K}$-equivalent</u>, <u>$\mathcal{C}$-equivalent</u>) if there exists a contact-equivalence (respectively $\mathcal{K}$-equivalence, $\mathcal{C}$-equivalence) (h,H) such that $(1, f_1) \circ h = H \circ (1, f_0)$.

We have, of course, already used the notion of equivalence extensively – in particular, in §0 we said heuristically that a versal parametrized unfolding of a map-germ f 'contains' representatives of all 'nearby' equivalence classes to that of f.

Now, as we pointed out in §0, any unfolding of f gives rise (non-uniquely) to a parametrized unfolding. In particular, then, a stable (equivalently, from §3, versal) unfolding does so, but of course the parametrized unfolding so obtained need not be versal in the category of parametrized unfoldings.

However, what is true is that such a parametrized unfolding 'contains' representatives of all 'nearby' contact-equivalence classes to that of f - more precisely, we should set up a category of parametrized unfoldings and 'contact-morphisms' and show that in this category an unfolding is versal if and only if it is stable as a map-germ.

That this is true seems as non-obvious as the equivalence of stability and versality for (non-parametrized) unfoldings, and it is equally difficult to offer any suggestion as to why it should be true other than a detailed proof of the type of those given in §3. We shall not give such a proof here, since the notion of 'contact-versality' in this form is not very useful for the purposes of this book. Instead, we shall obtain, in §6, a different version of the 'fact' that a parametrized unfolding of f which is stable as a map-germ 'contains' all 'nearby' contact-equivalence classes to that of f, via a transversality theorem.

Instead, let us use the discussion above as motivation: the close relationship outlined here between equivalence and contact-equivalence in the presence of stability suggests the following :

<u>Theorem</u> (4.3)

Let F^0, F^1 be stable unfoldings of the same dimension of germs f^0, f^1, where f^0, f^1 are contact-equivalent.

The F^0, F^1 are equivalent.

which has the important consequence:

<u>Corollary</u> (4.4)

If f^0, f^1 are contact-equivalent stable map-germs, then f^0, f^1 are equivalent.

<u>Proof</u>

f^0, f^1 are zero-dimensional stable unfoldings of themselves; so the result follows at once from (4.3).

In order to prove the theorem, we will need to relate contact-equivalence to the algebra of vector fields which we have been using to study unfoldings. The connecting link is provided by :

<u>Lemma</u> (4.5)

Map-germs $f,f' : (N,x_0) \to (P,y_0)$ are $\mathcal{C}$-equivalent if and only if $f^*m_P.C_N = f'^*m_P.C_N$.

<u>Proof</u>

'Only if' : Suppose f,f' are $\mathcal{C}$-equivalent, so that there exists a $\mathcal{C}$-equivalence $(1_N,H)$ such that $(1,f') \circ 1_N = H \circ (1,f)$.

Now $H(x,y_0) = (x,y_0)$ for all $x \in (N,x_0)$ (from the diagram of (4.2) a)), so that $H^*(p_P^*m_P . C_{N\times P}) \subset p_P^*m_P . C_{N\times P}$.

But H is a diffeomorphism, so H^* is an isomorphism, and thus $H^*(p_P^*m_P . C_{N\times P}) = p_P^*m_P . C_{N\times P}$.

So

$$f'^*m_P.C_N = (1,f')^*(p_P^*m_P . C_{N\times P}).C_N = (1,f)^*(H^*p_P^*m_P . C_{N\times P})).C_N$$

$$= (1,f)^*(p_P^*m_P . C_{N\times P}).C_N = f^*m_P.C_N .$$

'If': Suppose $f^*m_P.C_N = f'^*m_P.C_N$.

Let $x_1,\dots,x_n$ be a system of local co-ordinates at (N,x_0) , $y_1,\dots,y_p$ a system of local co-ordinates at (P,y_0) . Then there exist $v_{ij},w_{ij} \in C_N$ $(1 \le i,j \le p)$ such that

$$y_i \circ f = \sum_{j=1}^{p} v_{ij}.(y_j \circ f')$$

$$y_i \circ f' = \sum_{j=1}^{p} w_{ij}.(y_j \circ f)$$

For each $x \in (N,x_0)$, we define real $(p \times p)$-matrices $V_x = [v_{ij}(x)]$, $W_x = [w_{ij}(x)]$.

Let C be a real $(p \times p)$-matrix such that $C(I - V_{x_0}W_{x_0}) + W_{x_0}$ is invertible (such C exists: for, in general, given endomorphisms a,b of a vector space V, then there exists an endomorphism c such that $ca + b$ is invertible if (and only if) $\operatorname{Ker} a \cap \operatorname{Ker} b = \phi$ - just choose c to be any endomorphism of V such that $\operatorname{Im} c = c(a(\operatorname{Ker} b))$ and such that $\operatorname{Im} c \oplus \operatorname{Im} b = V$. In our particular case, if $W_{x_0}\alpha = 0$, then $(I - V_{x_0}W_{x_0})\alpha = \alpha$, so the condition of non-intersecting kernels is satisfied).

Now let $w'_{ij} = \sum_{k=1}^{p} c_{ik}(\delta_{kj} - \sum_{\ell=1}^{p} v_{k\ell}w_{\ell j}) + w_{ij} \in C_N$.

Then the matrix $W'_x = [w'_{ij}(x)]$ is invertible for $x = x_0$, and hence for x near x_0 .

Also, $y_i \circ f' = \sum_{j=1}^{p} w'_{ij}(y_j \circ f)$; and so $(1_N,H) \in \mathcal{C}$ defined by

$$\begin{cases} x_i \circ H = x_i & (i = 1,\dots,n) \\ y_i \circ H = \sum_{j=1}^{p} w'_{ij}(x_1,\dots,x_n).y_j & (i = 1,\dots,p) \end{cases}$$

is such that $(1,f') = H \circ (1,f)$.

Corollary (4.6)

Map-germs $f_i : (N_i,x_i) \to (P_i,y_i)$ $(i = 0,1)$ are contact-equivalent if and only if $\dim P_0 = \dim P_1$ and there exists an isomorphism $C_{N_0} \to C_{N_1}$ carrying $f_0^*m_{P_0}.C_{N_0}$ to $f_1^*m_{P_1}.C_{N_1}$.

Proof

Of course the condition $\dim P_0 = \dim P_1$ is trivial; and indeed we may as well assume $P_0 = P_1$ (=P, say), since if $k : (P_0,y_0) \to (P_1,y_1)$ is a germ of diffeomorphism, then $k^*m_{P_1} = m_{P_0}$.

'Only if' : Let (h,H) be a contact-equivalence such that $(1,f_1) \circ h = H \circ (1,f_0)$. Then $(1,f_1 \circ h) = ((h^{-1} \times 1_P) \circ H) \circ (1,f_0)$, so that $f_1 \circ h, f_0$ are $\mathcal{C}$-equivalent. Thus, by (4.5), $(h^{-1})^*$ provides the required isomorphism $C_{N_0} \to C_{N_1}$.

'If' : Suppose $\phi : C_{N_0} \to C_{N_1}$ is an isomorphism carrying $f_0^*m_P.C_{N_0}$ to $f_1^*m_P.C_{N_1}$.

Let $a_1,\dots,a_n$ be local co-ordinates at (N_0,x_0) , $b_1,\dots,b_n$ local co-ordinates at (N_1,x_1) . Then we may write

$$\phi^{-1}(b_i) = \sum_{j=1}^{n} h_{ij}a_j \quad (i = 1,\dots,n) .$$

Since ϕ (and ϕ^{-1}) are isomorphisms, the matrix $H_b = [h_{ij}(b)]$ is invertible for $b = x_0$ and hence for b near x_0, and so we may define a diffeomorphism $h : (N_0,x_0) \to (N_1,x_1)$ by

$$b_i \circ h = \sum_{j=1}^{n} h_{ij}.a_j \qquad (i=1,\ldots,n)$$

Then $\phi^{-1} = h^*$, so that $(f_1 \circ h)^* m_P C_{N_0} = f_0^* m_P . C_{N_0}$.

Thus by (4.5) $f_1 \circ h, f_0$ are $\mathcal{C}$-equivalent.

So f_0, f_1 are indeed contact-equivalent.

This leads to the geometrical result;

(4.7) Lemma

Let (F_k,i_k,j_k) $(k=0,1)$ be unfoldings of the same dimension of map-germs f_k.

Then F_0,F_1 are contact-equivalent germs if and only if f_0,f_1 are.

Proof

First note that, via composing with a germ of diffeomorphism $k : (P_0',j_0(P_0),y_0') \to (P_1',j_1(P_1),y_1')$, we may assume $P_0 = P_1$ (= P say), $P_0' = P_1' =$ (= P' say), $j_0 = j_1$ (= j say).

Now let us observe that, in general, if (F,i,j) is an unfolding of f, then

$$i^*(F^* m_{P'}.C_{N'}) = (i^*F^* m_{P'}).C_N \qquad \text{(since } i^*C_{N'} = C_N)$$

$$= (f^*j^* m_{P'}).C_N = f^* m_P.C_N \qquad \text{(since } j^* m_{P'} = m_P) .$$

Now for the proof :

'If' : Let (h,H) be a contact-equivalence such that $(1,f_1) \circ h = H \circ (1,f_0)$; so certainly $(f_1 \circ h)^* m_P.C_{N_0} = f_0^* m_P.C_{N_1}$.

Let $\tilde{h}$ be a germ of diffeomorphism $(N_0',x_0') \to (N_1',x_1')$ such that $\tilde{h} \circ i_0 = i_1 \circ h$. Then

$$i_0^*[(F_1 \circ \tilde{h})^* m_{P'}.C_{N_0'}] = (f_1 \circ h)^* m_P.C_{N_0} = f_0^* m_P.C_{N_0} = i_0^*[F_0^* m_{P'}.C_{N_0'}] .$$

But $\operatorname{Ker} i_0^* \subset F_0^* m_{P'}.C_{N_0'}$ (for $\operatorname{Ker} i_0^* = F_0^*(\operatorname{Ker} j_0^*)$), and so

$$(F_1 \circ \tilde{h})^* m_{P'} . C_{N_0'} = F_0^* m_{P'} . C_{N_0'} .$$

Thus $F_0, F_1 \circ \tilde{h}$ are $\mathcal{C}$-equivalent, and so F_0, F_1 are contact-equivalent.

'Only if' : Suppose F_0, F_1 are contact-equivalent; so that there is an isomorphism $\phi : C_{N_0} \to C_{N_1}$ carrying $F_0^* m_{P'} . C_{N_0'}$ onto $F_1^* m_{P'} . C_{N_1'}$.

Let us first prove the result in the case where rank $(df_0)_{x_0}$ (= rank $(df_1)_{x_1}$) = 0 .

In this case, Ker $i_0^* - m_{N_0'}^2 = F_0^* m_{P'} . C_{N_0'} - m_{N_0'}^2$.

Let $u_1, \ldots, u_k$ be local co-ordinates in N_0' such that Im $i_0 = \{u_1 = \ldots = u_k = 0\}$ and such that they extend to a system of local co-ordinates $u_1, \ldots, u_k, x_1, \ldots, x_n$ at (N_0', x_0') .

We may write (non-uniquely) $\phi(u_i) = v_i + a_i$, where $v_i \in$ Ker i_1^* , $a_i \in F_1^* m_{P'} . C_{N_1'} \cap m_{N_1'}^2$.

Then define a homomorphism $\phi' : C_{N_0'} \to C_{N_1'}$ by

$$\begin{cases} \phi'(u_i) = v_i & (i = 1, \ldots, k) \\ \phi'(x_j) = \phi(x_j) & (j = 1, \ldots, n) . \end{cases}$$

This is an isomorphism (for the induced map $m_{N_0'}/m_{N_0'}^2 \to m_{N_1'}/m_{N_1'}^2$ is the same as for ϕ) and it carries $F_0^* m_{P'} . C_{N_0'}$ into, hence onto, $F_1^* m_{P'} . C_{N_1'}$ (for $(\phi - \phi') m_{N_0'} \subset F_1^* m_{P'} C_{N_1'}$). Moreover, it carries Ker i_0^* onto Ker i_1^* ; and thus induces an isomorphism

$$i_0^*(F_0^* m_{P'} C_{N_0'}) \to i_1^*(F_1^* m_{P'} C_{N_1'})$$

i.e. an isomorphism $f_0^* m_P . C_{N_0} \to f_1^* m_P . C_{N_1}$.

Thus f_0, f_1 are contact-equivalent.

Now let us consider the case where rank $(df_0)_{x_0}$ (= rank $(df_1)_{x_1}$) > 0 : let $(\tilde{P}_0, y_0)$ be a submanifold-germ of (P, y_0) of codimension equal to rank $(df_0)_{x_0}$ such that f_0 is transverse to $\tilde{P}_0$. Let $\tilde{N}_0 = f_0^{-1}(\tilde{P}_0)$; and let $\tilde{f}_0 : (\tilde{N}_0, x_0) \to (\tilde{P}_0, y_0)$ be the restriction of f_0 . Clearly, then, $\tilde{f}_0$ is of rank 0 at x_0 ; also, $(f_0, \tilde{i}_0, \tilde{j}_0)$ is an unfolding of $\tilde{f}_0$ (where $\tilde{i}_0 : \tilde{N}_0 \to N_0, \tilde{j}_0 : \tilde{P}_0 \to P$ are the inclusions), as also is $(F_0, i_0 \circ \tilde{i}_0, j_0 \circ \tilde{j}_0)$.

Similarly, we can construct $\tilde{f}_1 : (\tilde{N}_1,x_1) \to (\tilde{P}_1,y_1)$ which is unfolded by f_1 and F_1.

Then, since F_0,F_1 are contact-equivalent, so are $\tilde{f}_0,\tilde{f}_1$, by the foregoing analysis. Hence, by the 'If" part of the lemma, f_0,f_1 are contact-equivalent.

Now, to connect with our algebra of vector fields, we have:

(4.8) Lemma

Let (h,H) be a $\mathcal{K}$-equivalence between map-germs f,f'; and let $e_{(h,H)} : \theta_f \to \theta_{f'}$ be the C_N-module isomorphism defined by $(0,e_{(h,H)}\phi) = dH \circ (0,\phi) \circ h^{-1}$ $(\phi \in \theta_f)$ (where we identify $\theta_{(1,f)} \equiv \theta_N \oplus \theta_f$).

Then $e_{(h,H)}$ induces an $\mathbb{R}$-vector space isomorphism

$$\varepsilon_{(h,H)} : \mathcal{N}_f \to \mathcal{N}_{f'}$$

(where $\mathcal{N}_f = \theta_f / tf(\theta_N) + f^*m_P.\theta_f$).

Proof

We may write $e_{(h,H)}$ as the composite $e_{(h,h\times 1_P)} \circ e_{(1_N,H')}$, where $H' = (h^{-1} \times 1_P) \circ H$, and so it is sufficient to show that each of $\varepsilon_{(h,h\times 1_P)}$, $\varepsilon_{(1_N,H)}$ is well-defined.

First, then, the case $H = h \times 1_P$.

We have $f' = f \circ h^{-1}$, $e_{(h,h\times 1_P)} = wh^{-1}$.

Thus $e_{(h,h\times 1_P)}(f^*m_P.\theta_f) = (h^{-1})^*f^*m_P.\theta_{f'} = f'^*m_P.\theta_{f'}$.

Also, if $\xi \in \theta_N$, then

$$\begin{aligned} wh^{-1}(tf(\xi)) &= wh^{-1} \circ Tf' \circ dh \circ \xi = Tf' \circ dh \circ \xi \circ h^{-1} \\ &= tf'(Th \circ \xi \circ h^{-1}). \end{aligned}$$

Thus $e_{(h,h\times 1_P)}(tf(\theta_N)) = tf'(\theta_N)$.

So $\varepsilon_{(h,h\times 1_P)}$ is indeed well-defined.

Now let us consider the case $h = 1_N$.

By (4.5), $f^*m_P.C_N = f'^*m_P.C_N$, so that, since $e_{(1,H)}$ is just a restriction of the homomorphism $dH : \theta_{(1,f)} \to \theta_{(1,f')}$ over H^*, we have

$$e_{(1,H)}(f^*m_P.\theta_f) = f'^*m_P.\theta_{f'}.$$

Also, if $\xi \in \theta_N$, then

$$(0,e_{(1,H)}(tf(\xi)) = dH(0,Tf\circ\xi) = TH\circ T(1,f)\circ\xi - TH\circ(\xi,0)\circ(1,f)$$
$$= T(1,f')\circ\xi - TH\circ(\xi,0)\circ(1,f)\ .$$

Now $TH\circ(\xi,0)\circ(1,f) = (\xi,\phi)$, where $\phi \in f'^*m_P.\theta_{f'}$, because $H(x,y_0) = (x,y_0)$ for all $x \in (N,x_0)$.

Thus we have $e_{(1,H)}(tf(\xi)) = tf'(\xi) + \phi$, and so we have shown

$$e_{(1,H)}(tf(\theta_N)) \subset tf'(\theta_N) + f'^*m_P.\theta_f\ .$$

Thus, also, $\varepsilon_{(1,H)}$ is well-defined.

We can now give

Proof of Theorem (4.3)

First of all, let us observe that it is sufficient to consider the case $rk(df^0_{x_0}) = rk(df^1_{x_1}) = 0$ (for as in Lemma (4.7), f^0,f^1 are unfoldings of rank 0 germs $\tilde{f}^0,\tilde{f}^1$ which, by that lemma are contact-equivalent since f^0,f^1 are; and of course F^0,F^1 are stable unfoldings of $\tilde{f}^0,\tilde{f}^1$).

Next, let us remark that it is sufficient to consider the case when f^0,f^1 are $\mathcal{C}$-equivalent, with the $\mathcal{C}$-equivalence between them such that $H(x_o,y) = y$ for all $y \in (P,y_0)$
(for if $(1,f^1)\circ h = H\circ(1,f^0)$, then $(1,(H^{x_0})^{-1}\circ f^1\circ h) = (h^{-1},h^{-1}\times(H^{x_0})^{-1})\circ H\circ(1,f)$ (where $H^{x_0} : (P_0,y_0) \to (P_1,y_1)$ is defined by $H^{x_0}(y) = H(x_0,y))$, so that we can replace f^1 by the equivalent germ $(H^{x_0})^{-1}\circ f^1\circ h$, which is $\mathcal{C}$-equivalent to f^0, and also replace F^1 by an equivalent germ $\tilde{k}^{-1}\circ F^1\circ\tilde{h}$, where $\tilde{h} : (N_0',x_0') \to (N_1',x_1')$, $\tilde{k} : (P_0',y_0') \to (P_1',y_1')$ are germs of diffeomorphism such that $\tilde{h}\circ i_0 = i_1\circ h$, $\tilde{k}\circ j_0 = j_1\circ H^{x_0}$).

Let us also observe that, by Theorem (3.1), it will be enough to show that any stable unfolding of f^0 (of the relevant dimension, k say) is equivalent to any stable unfolding of f^1 of dimension k.

Now, in general, if (F,i,j) is any unfolding of a map-germ f, we have
$(\rho_{F,f})^{-1}m_N\mathcal{N}_f = \text{Im}\ TF_{x_0'}$
(For $\eta \in \theta_F$ is such that $\eta_{y_0'} \in \text{Im}\ TF_{x_0'} \Longleftrightarrow$ there exists $\xi \in \theta_{N'}$ such that

$$wF(\eta) - tF(\xi) \in m_{N'} \cdot \theta_F \Longleftrightarrow \rho_F(\eta_{y_0'}) \in m_{N'} \mathcal{N}_F$$

$$\Longleftrightarrow \rho_{F,f}(\eta_{y_0'}) = (q_{F,f})^{-1} \rho_F(\eta_{y_0}) \in m_N \mathcal{N}_F\ ).$$

Also, we have $\mathcal{N}_f = m_N \mathcal{N}_f + [wf(TP)]$, and $[wf(TP)] = \rho_{F,f}(\mathrm{Im}(Tj_{y_0}))$.

So, since F is a stable unfolding of f if and only if $\rho_{F,f}$ is surjective (by (2.6)), it follows that F is stable if and only if the restriction of $\rho_{F,f}$

$$\bar{\rho}_{F,f} : \mathrm{Im}\ TF_{x_0'} \to m_N f$$

is surjective.

Now let us consider our $\mathcal{C}$-equivalent map-germs f^0, f^1 , which are of rank 0 at x_0 .

Let U be a neighbourhood of 0 in $\mathbb{R}^k$, and let $g^0 : (N \times U, x_0 \times 0) \to (P, y_0)$ be a map-germ s.t. $g^0(x,0) = f^0(x)$ and such that $\{[\frac{\partial}{\partial u_i} \cdot g^0]\}$ span $m_N \mathcal{N}_f$ (Such g^0 exists by (1.9); it is possible to span $m_N \mathcal{N}_f$ with k elements, since f^0 does have a stable unfolding of dimension k (and hence rank k at x_0'). Then the germ $F^0 : (N \times U, x_0 \times 0) \to (P \times U, y_0 \times 0)$ defined by $F^0(x,u) = (g^0(x,u), u)$ is a stable unfolding of f^0 (for, as in (2.8), $\rho_{F^0,f^0}(\frac{\partial}{\partial u_i}\Big|_{y_0 \times 0}) = -[\frac{\partial}{\partial u_i} \cdot g^0]$).

Now let $g^1 : (N \times U, x_0 \times 0) \to (P, y_0)$ be defined by $(x, g^1(x,u)) = H(x, g^0(x,u))$ (where $(1_N, H)$ is the $\mathcal{C}$-equivalence between f^0 and f^1); then clearly $g^1(x,0) = f^1(x)$, and $e_{(1,H)}(\frac{\partial}{\partial u_i} \cdot g^0) = \frac{\partial}{\partial u_i} \cdot g^1$. Thus, if we define $F^1 : (N \times U, x_0 \times 0) \to (P \times U, y_0 \times 0)$ by $F^1(x,u) = (g^1(x,u), u)$, then F^1 is a stable unfolding of f^1 . Thus it remains to show that F^0, F^1 are equivalent germs. This we will do by showing that they are part of an $(\mathbb{R}, I)$-family of stable germs trivial over $(\mathbb{R}, I)$.

Let $\{x_1, \ldots, x_n\}$ be a system of local co-ordinates at (N, x_0) , and let $\{y_1, \ldots, y_p\}$ be a system of local co-ordinates at (P, y_0) .

Define $F^a : (N \times U, x_0 \times 0) \to (P \times U, y_0 \times 0)$ $(a \in (\mathbb{R}, I))$ by

$$\begin{cases} y_i \circ F^a = (1-a) \cdot y_i \circ F^0 + a \cdot y_i \circ F^1 \\ u_i \circ F^a = u_i \end{cases}$$

so that F^a is an unfolding of $f^a : (N,x_0) \to (P,y_0)$ defined by

$$y_i \circ f^a = (1-a) \cdot y_i \circ f^0 + a \cdot y_i \circ f^1 .$$

Indeed $(1,f^a) = H^a \circ (1,f^0)$, where $H^a : (N \times P, x_0 \times y_0) \to (N \times P, x_0 \times y_0)$ is defined by

$$x_i \circ H^a = x_i$$

$$y_i \circ H^a = (1-a)y_i \circ 1_{N\times P} + a y_i \circ H$$

so that $(1_N, H^a)$ is a $\mathcal{C}$-equivalence.

Of course F^a is of the form $F^a(x,u) = (g^a(x,u),u)$, where

$$y_i \circ g^a = (1-a)y_i \circ g^0 + a \cdot y_i \circ g^1$$

and clearly we have $e_{(1,H^a)}(\frac{\partial}{\partial u_i} \cdot g^0) = \frac{\partial}{\partial u_i} \cdot g^a$.

Hence

$$\rho_{F^a,f^a}(\frac{\partial}{\partial u_i}\Big|_{y_0\times 0}) = -[\frac{\partial}{\partial u_i} \cdot g^a] = -[e_{(1,H^a)}(\frac{\partial}{\partial u_i} \cdot g^0)]$$

$$= \varepsilon_{(1,H^a)} \bar{\rho}_{F^0,f^0}(\frac{\partial}{\partial u_i}\Big|_{y_0\times 0}) .$$

Thus $\bar{\rho}_{F^a,f^a}$ is surjective, so that F^a is a stable unfolding of f^a .

Now let $G : (N \times U \times \mathbb{R}, x_0 \times 0 \times I) \to (P \times U \times \mathbb{R}, y_0 \times 0 \times I)$ be defined by $G(x,u,a) = (F^a(x,u),a)$. We aim to show that G is trivial over $(\mathbb{R},I)$; that is, that there exist retractions r,s to $1_{N\times U} \times 0$, $1_{P\times U} \times 0$ respectively such that the diagram

$$\begin{array}{ccc}
(N \times U, x_0 \times 0) & \xrightarrow{\quad F^0 \quad} & (P \times U, y_0 \times 0) \\
r \uparrow & & \uparrow s \\
(N \times U \times \mathbb{R}, x_0 \times 0 \times I) & \xrightarrow{\quad G \quad} & (P \times U \times \mathbb{R}, y_0 \times 0 \times I) \\
& \searrow \qquad \swarrow \pi_{\mathbb{R}} & \\
& (\mathbb{R}, I) &
\end{array}$$

commutes. If we show this, then $F^1 = s^1 \circ F^0 \circ (r^1)^{-1}$, where the diffeomorphism-germs r^1 of $(N \times U, x_0 \times 0)$, s^1 of $(P \times U, y_0 \times 0)$ are given by $r^1(x,u) = r(x,u,1)$, $s^1(y,u) = s(y,u,1)$, so that F^0, F^1 will indeed be equivalent.

As in (3.2), to construct the desired retractions r,s it will be sufficient to find, for each $a \in (\mathbb{R},I)$, elements $\xi'_a \in p^*_{N\times U} m_{N\times U} \theta_{P_{N\times U}}$, $\eta'_a \in p^*_{P\times U} m_{N\times U} \cdot \theta_{P\times U}$

such that $\xi_a = (\xi'_a, \frac{\partial}{\partial u} \circ \pi_{\mathbb{R}} \circ G_a)$ is a lift for $\eta_a = (\eta'_a, \frac{\partial}{\partial u} \circ \pi_{\mathbb{R}})$ over G_a.

(Here $G_a : (N \times U \times \mathbb{R}, x_0 \times 0 \times a) \to (P \times U \times \mathbb{R}, y_0 \times 0 \times a)$ is the germ-restriction of G; and we have identified

$$\theta_{N\times U\times\mathbb{R}} = \theta_{p_{N\times U}} \oplus \theta_{\pi_{\mathbb{R}} \circ G_a}, \quad \theta_{P\times U\times\mathbb{R}} = \theta_{p_{P\times U}} \oplus \theta_{\pi_{\mathbb{R}}}).$$

To do this, we proceed as follows:

$$\text{let} \quad \phi_a = tG_a(0, \frac{\partial}{\partial u} \circ \pi_{\mathbb{R}} \circ G_a) - wG_a(0, \frac{\partial}{\partial u} \circ \pi_{\mathbb{R}}).$$

Then $\phi_a = \sum_{i=1}^{p} p^*_{N\times U}(y_i \circ F^1 - y_i \circ F^0) . wG_a(\frac{\partial}{\partial y_i} \circ p_{P\times U}, 0)$.

Now F^a is $\mathcal{C}$-equivalent to F^0 (the $\mathcal{C}$-equivalence being (essentially) $H^a \times 1_U \times 1_U$), so that

$$F^{a*}m_{P\times U} . C_{N\times U} = F^{0*}m_{P\times U} . C_{N\times U}.$$

Thus

$$G_a{}^* p^*_{P\times U} m_{P\times U} . C_{N\times U\times\mathbb{R}} = p^*_{N\times U} F^{0*} m_{P\times U} . C_{N\times U\times\mathbb{R}}.$$

Now
$$y_i \circ F^1 \in F^{1*}m_{P\times U} . C_{N\times U} = F^{0*}m_{P\times U} . C_{N\times U}.$$

So

$$p^*_{N\times U}(y_i \circ F^1 - y_i \circ F^0) \in p^*_{N\times U} F^{0*} m_{P\times U} . C_{N\times U\times\mathbb{R}} = G_a{}^* p^*_{P\times U} m_{P\times U} . C_{N\times U\times\mathbb{R}}$$

and thus

$$\phi_a \in G_a{}^* p^*_{P\times U} m_{P\times U} . \theta_{G_a, \pi_{\mathbb{R}}}.$$

Now, by (2.3), we have

$$\theta_{G_a, \pi_{\mathbb{R}}} = tG_a(\theta_{N\times U\times\mathbb{R}}, \pi_{\mathbb{R}} \circ G_a) + wG_a(\theta_{P\times U\times\mathbb{R}}, \pi_{\mathbb{R}})$$

so that

$$G_a{}^* p^*_{P\times U} m_{P\times U} . \theta_{G_a, \pi_{\mathbb{R}}} \subset tG_a(p^*_{N\times U} m_{N\times U} . \theta_{N\times U\times\mathbb{R}, \pi_{\mathbb{R}} \circ G_a}) + wG_a(p^*_{P\times U} m_{P\times U} . \theta_{P\times U\times\mathbb{R}, \pi_{\mathbb{R}}})$$

Hence, via the natural identifications

$$\theta_{p_{N\times U}} = \theta_{N\times U\times\mathbb{R}, \pi_{\mathbb{R}} \circ G_a}, \quad \theta_{p_{P\times U}} = \theta_{P\times U\times\mathbb{R}, \pi_{\mathbb{R}}}$$

we can find $\xi'_a \in p^*_{N\times U} m_{N\times U} . \theta_{p_{N\times U}}$, $\eta'_a \in p^*_{P\times U} m_{P\times U} \theta_{p_{P\times U}}$

such that
$$\phi_a = tG_a(\xi'_a, 0) - wG_a(\eta'_a, 0).$$

So, since $\phi_a = tG_a(0, \frac{\partial}{\partial u} \circ \pi_{\mathbb{R}} \circ G_a) - wG_a(0, \frac{\partial}{\partial u} \circ \pi_{\mathbb{R}})$, we have $tG_a(\xi'_a, \frac{\partial}{\partial u} \circ \pi_{\mathbb{R}} \circ G_a) = wG_a(\eta'_a, \frac{\partial}{\partial u} \circ \pi_{\mathbb{R}})$, so that $\xi_a = (\xi'_a, \frac{\partial}{\partial u} \circ \pi_{\mathbb{R}} \circ G_a)$ is indeed a lift for $\eta_a = (\eta'_a, \frac{\partial}{\partial u} \circ \pi_{\mathbb{R}})$ over G_a, as required, and the proof is complete.

§5. Determinacy

We now turn our attention to the question of whether a map germ is equivalent (or contact-equivalent) to a polynomial.

(5.1) **Definition**

Map-germs $f,f' : (N,x_0) \to (P,y_0)$ are r-jet equivalent if $(f^* - f'^*)m_P \subset m_N^{r+1}$.

This is clearly an equivalence relation; we call the equivalence class of f $j^r f$.

A more usual definition of r-jet, in terms of partial derivatives, is equivalent to that given above, as is shown in the following:

(5.2) **Lemma**

Let $f,f' : (N,x_0) \to (P,y_0)$ be map-germs.

Then $j^r f = j^r f'$ if and only if the partial derivatives of f and f' (with respect to some, and hence any, systems of local co-ordinates) agree for all orders $\leq r$.

To prove this, we will require, first

(5.3) **Sublemma**

Let $\phi \in C_{N,x_0}$. Then $\phi \in m_N^{r+1}$ if and only if all partial derivatives of ϕ, with respect to any system of local co-ordinates for (N,x_0), of orders $\leq r$ vanish.

Proof

Let $x_1,\ldots,x_n$ be a system of local co-ordinates for (N,x_0).

If $r = 0$, the result holds (by definition!).

Otherwise let us assume inductively that the result holds for jets of order $< r$.

If $\phi \in m_N^{r+1}$, then $\phi = \sum_{i=1}^{n} \phi_i . x_i$, with $\phi_i \in m_N^r$ (since $x_1,\ldots,x_n$ generate m_N, by (1.2)). By our inductive hypothesis, then, the partial derivatives of ϕ_i vanish for all orders $\leq r-1$, so that the partial derivatives of $x_i . \phi_i$, and so ϕ, vanish for orders $\leq r$ (differentiation of products!).

Conversely, if the partial derivatives of ϕ vanish for all orders $\leq r$, let ϕ_i be the function defined by $\phi_i(x) = \int_0^1 \frac{\partial\phi}{\partial x}(0,\ldots,tx_i,x_{i+1},\ldots,x_n)\cdot dt$. Then (as in (1.2)) $\phi = \sum_{i=1}^{n} \phi_i . x_i$. Now $x_i . \phi_i(x) = \phi(0,\ldots,x_i,\ldots,x_n) - \phi(0,\ldots,x_{i+1},\ldots,x_n)$ so that, since the partial derivatices of ϕ for orders $\leq r$ vanish, so do those of $x_i . \phi_i(x)$. Then it is easy to see that the partial derivatives of ϕ_i must vanish for orders $\leq r-1$; so by our inductive hypothesis $\phi_i \in m_N^r$; whence $\phi \in m_N^{r+1}$.

So the inductive step, and hence the proof, is complete.

Now we have ;

Proof of (5.2)

All partial derivatives of f,f' of orders $\leq r$ agree if and only if the same is true for $v \circ f, v \circ f'$ for any $v \in m_P$; which, by (5.1), is true if and only if $v \circ f - v \circ f' = (f^* - f'^*)(v) \in m_N^{r+1}$ for any $v \in m_P$.

We can derive further consequences of jet-equivalence as follows: let $y_1,\ldots,y_p$ be a system of local co-ordinates at (P,y_0). Then $wf(\frac{\partial}{\partial y_1}),\ldots,wf(\frac{\partial}{\partial y_p})$ is a free basis for θ_f as a C_N-module. We will identify θ_f and $\theta_{f'}$ in the following lemma by $wf(\frac{\partial}{\partial y_i}) \leftrightarrow wf'(\frac{\partial}{\partial y_i})$.

(5.4) Lemma

Let $f,f' : (N,x_0) \to (P,y_0)$ be such that $j^r f = j^r f'$.

Then

a) $$tf'(\theta_N) \subset tf(\theta_N) + m_N^r \theta_f$$

b) $$f'^* m_P \theta_{f'} \subset (f^* m_P + m_N^{r+1})\theta_f .$$

Proof

a) Let $x_1,\ldots,x_n$ be a system of local co-ordinates at (N,x_0).

Then, since $\frac{\partial}{\partial x_1},\ldots,\frac{\partial}{\partial x_n}$ is a free basis for the C_N-module θ_N, we can write $\xi = \sum_{i=1}^{n} \xi_i . \frac{\partial}{\partial x_i}$ $(\xi_i \in C_N)$ for any $\xi \in \theta_N$.

Thus $$tf(\xi) = \sum_{i=1}^{n} \xi_i . tf(\frac{\partial}{\partial x_i}) = \sum_{i,j} \xi_i \frac{\partial}{\partial x_i}(f^*(y_j)).wf(\frac{\partial}{\partial y_j}) ,$$

and so $\quad tf'(\xi) = tf(\xi) + \sum_{i,j} \xi_i \frac{\partial}{\partial x_i}(f'^*(y_j) - f^*(y_j)).wf(\frac{\partial}{\partial y_j})$.

Since f, f' have the same r-jet, all partial derivatives of $f'^*(y_j) - f^*(y_j)$ of orders $\leq r$ vanish, and so all partial derivatives of $\frac{\partial}{\partial x_i}(f'^*(y_j) - f^*(y_j))$ of orders $\leq r-1$ vanish.

So, by (5.1), $\frac{\partial}{\partial x_i}(f^*(y_j) - f'^*(y_j)) \in m_N^r$, and thus $tf'(\xi) - tf(\xi) \in m_N^r \theta_f$.

b) Since f, f' have the same r-jet, $f'^* m_P \subset f^* m_P + m_N^{r+1}$, and the result follows at once.

(5.5) Definition

A map-germ $f : (N, x_0) \to (P, y_0)$ is $\mathcal{K}$-determined by its r-jet (respectively determined by its r-jet) if every map-germ $f' : (N, x_0) \to (P, y_0)$ such that $j^r f = j^r f'$ is $\mathcal{K}$-equivalent (respectively equivalent) to f .

Thus, if f is determined by its r-jet, it is equivalent to a rather particular polynomial - its Taylor series (with respect to any local co-ordinates) truncated after order r . However, determinacy is rather hard to calculate; but for stable map-germs can clearly be arrived at via $\mathcal{K}$-determinacy (using (4.4)), for which we now give a sufficient condition:

(5.6) Theorem

Let $f : (N, x_0) \to (P, y_0)$ be a map-germ such that

$$m_N^r \theta_f \subset tf(\theta_N) + f^* m_P \theta_f \ .$$

Then f is $\mathcal{K}$-determined by its (r+1)-jet.

Proof

Let $f' : (N, x_0) \to (P, y_0)$ be a map-germ such that $j^{r+1} f = j^{r+1} f'$.

Let $(y_1, \ldots, y_p)$ be a system of local co-ordinates at (P, y_0) . For $a \in \mathbb{R}$, define $f^a : (N, x_0) \to (P, y_0)$ by

$$y_i \circ f^a = (1-a).y_i \circ f + a.y_i \circ f' \qquad (i=1,\ldots,p)$$

Then define $g : (N \times \mathbb{R}, x_0 \times I) \to (P, y_0)$ by $g(x,a) = f^a(x)$, and $\gamma : (N \times \mathbb{R}, x_0 \times I) \to (N \times P \times \mathbb{R}, x_0 \times y_0 \times I)$ by $\gamma(x,a) = (x, g(x,a), a)$.

We aim to find retractions r, S to $1_N \times 0, 1_{N\times P} \times 0$ respectively such that the following diagram commutes:

$$\begin{array}{ccccc}
(N,x_0) & \xrightarrow{(1,f)} & (N\times P, x_0\times y_0) & \longrightarrow & (N,x_0) \\
\uparrow r & & \uparrow S & & \uparrow r \\
(N\times \mathbb{R}, x_0 \times I) & \xrightarrow{\gamma} & (N\times P\times \mathbb{R}, x_0\times y_0\times I) & \xrightarrow{p_{N\times\mathbb{R}}} & (N\times\mathbb{R}, x_0\times I) \\
& \searrow^{\pi'_{\mathbb{R}}} & \downarrow \pi_{\mathbb{R}} & \swarrow^{\pi'_{\mathbb{R}}} & \\
& & (\mathbb{R}, I) & &
\end{array}$$

If we can do so, then $r^1 : (N,x_0) \to (N,x_0)$, defined by $r^1(x) = r(x,1)$, and $S^1 : (N\times P, x_0\times y_0) \to (N\times P, x_0\times y_0)$, defined by $S^1(x,y) = S(x,y,1)$, give a $\mathcal{K}$-equivalence (r^1, S^1) between f and f' , and the proof will be complete.

Let $g_a : (N\times\mathbb{R}, x_0\times a) \to (P,y_0)$, $\gamma_a : (N\times\mathbb{R}, x_0\times a) \to (N\times P\times\mathbb{R}, x_0\times y_0\times a)$ be the germ-restrictions of g,γ (for $a \in I$) .

We shall identify, in the natural way,

$$\theta_{\gamma_a} = \theta_{p_N} \oplus \theta_{g_a} \oplus \theta_{\pi_{\mathbb{R}}}$$

$$\theta_{N\times P\times\mathbb{R},\, x_0\times y_0\times a} = \theta_{p_N} \oplus \theta_{p_P} \oplus \theta_{\pi_{\mathbb{R}}}$$

$$\theta_{N\times\mathbb{R},\, x_0\times a} = \theta_{p'_N} \oplus \theta_{\pi'_{\mathbb{R}}}$$

(where p_N, p_P are the natural projections of $(N\times P\times\mathbb{R}, x_0\times y_0\times a)$ to $(N,x_0), (P,y_0)$ respectively, and p'_N is the natural projection $(N\times\mathbb{R}, x_0\times a) \to (N,x_0)$).

By similar arguments to those of (3.2), for the construction of r,S to be possible it will be sufficient to find, for each $a \in I$, elements $\xi'_a \in m_N\theta_{p'_N}$, $\eta'_a \in m_N m_P \theta_{p_P}$ such that $\xi_a = (\xi'_a, \frac{\partial}{\partial u}\circ\pi'_{\mathbb{R}})$ is a lift for $\eta_a = (\xi'_a \circ p_{N\times\mathbb{R}}, \eta'_a, \frac{\partial}{\partial u}\circ\pi_{\mathbb{R}})$ over γ_a .

Let $$\phi_a = tg_a((0, \tfrac{\partial}{\partial u}\circ\pi'_{\mathbb{R}})) \in \theta_{g_a} .$$

With respect to the system of local co-ordinates $(y_1,\ldots,y_p)$, we can write

$$\phi_a = \sum_{i=1}^{p} (\phi_a)_i\, wg_a\!\left(\frac{\partial}{\partial y_i}\right) \qquad (\phi_a)_i \in C_{N\times\mathbb{R},\, x_0\times a}$$

and it is easy to calculate that

$$(\phi_a)_i = y_i \circ f' - y_i \circ f \in m_N^{r+2} C_{N\times\mathbb{R}}$$

so that
$$\phi_a \in m_N^{r+2}\theta_{g_a} .$$

Now we claim that

$$m_N^r \theta_{g_a} \subset tg_a(\theta_{N\times\mathbb{R},\pi'_{\mathbb{R}}}) + g_a^* m_P . \theta_{g_a} .$$

To see this, we argue as follows :

from the hypothesis of the theorem, we have

$$m_N^r \theta_f \subset tf(\theta_N) + f^* m_P . \theta_f .$$

Since $j^{r+1}f^a = j^{r+1}f$, it follows, by applying (5.4), that

(*)
$$m_N^r \theta_{f^a} \subset tf^a(\theta_N) + (f^{a*} m_P + m_N^{r+1})\theta_{f^a} .$$

Now recall (from (2.2)) that there are natural isomorphisms

$$\theta_{g_a}/m_{\mathbb{R},a}\theta_{g_a} \cong \theta_{f^a}$$

(g_a gives rise in the obvious way to an unfolding F_a of f^a , and we can identify θ_{g_a} with $\theta_{F_a,\pi}$, where π is the projection $(P\times\mathbb{R}, y_0\times a) \to (\mathbb{R}, a)$)

$$\theta_{N\times\mathbb{R},\pi'_{\mathbb{R}}}/m_{\mathbb{R},a}\theta_{N\times\mathbb{R},\pi'_{\mathbb{R}}} \cong \theta_N .$$

So from (*) we can deduce

$$m_N^r \theta_{g_a} \subset tg_a(\theta_{N\times\mathbb{R},\pi'_{\mathbb{R}}}) + g_a^* m_P . \theta_{g_a} + m_N^{r+1}\theta_{g_a} .$$

Thus it follows from (1.10) that the C_N-module (via $\pi'^*_{\mathbb{R}}$)

$$\{m_N^r \theta_{g_a} + tg_a(\theta_{N\times\mathbb{R},\pi'_{\mathbb{R}}}) + g_a^* m_P . \theta_{g_a}\}/\{tg_a(\theta_{N\times\mathbb{R},\pi'_{\mathbb{R}}}) + g_a^* m_P . \theta_{g_a}\} = 0 ,$$

and the claim follows.

It certainly follows, then, that

$$\phi_a \in tg_a(m_N \theta_{N\times\mathbb{R},\pi'_{\mathbb{R}}}) + m_N . g_a^* m_P . \theta_{g_a}$$

so that there exist $\xi'_a \in m_N\theta_{N\times\mathbb{R},\pi'_{\mathbb{R}}}$, $\eta''_a \in m_N . g_a^* m_P . \theta_{g_a}$

such that
$$\phi_a = tg_a(-\xi'_a) + \eta''_a .$$

We will, as usual, identify $\theta_{N\times\mathbb{R},\pi'_{\mathbb{R}}}$ with $\theta_{p'_N}$, and so identify $\xi'_a \in m_N\theta_{N\times\mathbb{R},\pi'_{\mathbb{R}}}$ with $(\xi'_a, 0) \in \theta_{p'_N} \oplus \theta_{\pi'_{\mathbb{R}}}$.

Now γ_a is an immersion, so, since $g_a = p_P \circ \gamma_a$, there exists $\eta'_a \in m_N m_P \theta_{p_P}$ such that $\eta'_a \circ \gamma_a = \eta''_a$. It remains to show, then, that $\xi_a = (\xi'_a, \frac{\partial}{\partial u} \circ \pi'_{\mathbb{R}})$ is a lift for $\eta_a = (\xi'_a \circ p_{N\times\mathbb{R}}, \eta'_a, \frac{\partial}{\partial u} \circ \pi_{\mathbb{R}})$ over γ_a:

We have

$$\begin{aligned} T\gamma_a \circ \xi_a &= (\xi'_a \circ p_{N\times\mathbb{R}}, \; tg_a(\xi'_a) + tg_a(\tfrac{\partial}{\partial u} \circ \pi'_{\mathbb{R}}), \tfrac{\partial}{\partial u} \circ \pi_{\mathbb{R}}) \\ &= (\xi'_a \circ p_{N\times\mathbb{R}}, \; tg_a(\xi'_a) + \phi_a, \tfrac{\partial}{\partial u} \circ \pi_{\mathbb{R}}) \\ &= (\xi'_a \circ p_{N\times\mathbb{R}}, \; \eta''_a, \tfrac{\partial}{\partial u} \circ \pi_{\mathbb{R}}) = \eta_a \circ \gamma_a \end{aligned}$$

and so the proof is complete.

(5.7) Corollary

If f is of FST, then f is $\mathcal{K}$-determined by its $(\chi_f + 1)$-jet.

Proof

By hypothesis, $\chi_f = \dim_{\mathbb{R}} \{\theta_f / tf(\theta_N) + f^* m_P . \theta_f\} = r < \infty$. Thus, by (1.12), $m_N^r \theta_f \subset tf(\theta_N) + f^* m_P . \theta_f$, and so, by (5.6), f is $\mathcal{K}$-determined by $j^{r+1} f$.

(5.8) Corollary

If f is stable, then f is determined by its $(p + 1)$-jet (where $p = \dim. P$).

Proof

If f is stable, then $\rho_f : TP_{y_0} \to \theta_f / tf(\theta_N) + f^* m_P . \theta_f$ is surjective, so that $\dim_{\mathbb{R}} \{\theta_f / tf(\theta_N) + f^* m_P . \theta_f\} \le p$. Thus, by (1.12), $m_N^p \theta_f \subset tf(\theta_N) + f^* m_P . \theta_f$.

Now let f' be such that $j^{p+1} f' = j^{p+1} f$. Then, by (5.4), we can conclude that

$$tf'(\theta_N) + f'^* m_P . \theta_{f'} \equiv tf(\theta_N) + f^* m_P . \theta_f$$

so that $\mathcal{N}_f, \mathcal{N}_{f'}$ can be identified.

But then $\rho_f, \rho_{f'}$ can be identified (since these are induced respectively by wf, wf', and our identifications are given by $wf(\frac{\partial}{\partial y_i}) \leftrightarrow wf'(\frac{\partial}{\partial y_i})$), so that $\rho_{f'}$ is also surjective, and hence f' is stable.

But, by (5.6), it follows that f' is $\mathcal{K}$-equivalent to f, so that, by (4.4), f' is actually equivalent to f.

§6. Jet-spaces and a transversality theorem

In this section we will obtain the 'contact-transversality' theorem advertised in §4.

First of all, however, we will require some material on jet-spaces.

(6.1) The set $J^r(N,P)$ (for any integer $r \geq 1$) consists of all r-jets of germs $(N,x) \to (P,y)$, for any $x \in N$, $y \in P$.

There is an obvious projection $p^{r,0} : J^r(N,P) \to N \times P$ induced by the function assigning (x,y) to any map-germ $(N,x) \to (P,y)$. We write $J^r(N,P)_{x_0}$ for $(p^{r,0})^{-1}\{x_0\} \times \{P_0\}$, and $J^r(N,P)_{x_0,y_0}$ (or J^r) for $(p^{r,0})^{-1}\{x_0\} \times \{y_0\}$.

There are also projections $p^{r,s} : J^r(N,P) \to J^s(N,P)$ $(1 \leq s \leq r)$ given by $p^{r,s}(j^r f) = j^s f$.

$J^r(N,P)$ is topologized as a smooth manifold, in such a way that the $p^{r,s}(0 \leq s \leq r)$ are projections of smooth fibre-bundles, by giving the following local charts:

if $(x_1,\ldots,x_n)$ is a system of local co-ordinates defined on a neighbourhood U of x_0 in N, and if $(y_1,\ldots,y_p)$ is a system of local co-ordinates defined on a neighbourhood V of y_0 in P, then $(X_i,Y_j,Z_{k,\sigma})$, where $i=1,\ldots,n$, $j=1,\ldots,p$ and $k=1,\ldots,p$, $\sigma = (\sigma_1,\ldots,\sigma_n)$ with $1 \leq \sigma_1 + \ldots + \sigma_n \leq r$ and $\sigma_i \geq 0$, is a system of local co-ordinates at $0 \in J^r(N,P)_{x_0,y_0}$ (where 0 is the r-jet of the constant germ) defined on $(p^{r,0})^{-1}U \times V$, where, if $z \in (p^{r,0})^{-1}U \times V$ is the r-jet of a germ $f : (N,x) \to (P,y)$, then

$$X_i(z) = x_i(x)$$

$$Y_j(z) = y_j(y)$$

$$Z_{k,\sigma}(z) = \frac{\partial^{\sigma_1+\ldots+\sigma_n}(y_k \circ f)}{\partial x_1^{\sigma_1}\ldots\partial x_n^{\sigma_n}}$$

(That these functions are well-defined follows from (5.2); that they form a system of local co-ordinates becomes clear by considering germs of polynomial mappings).

(6.2) In fact, we can be rather more exact about the fibre-bundle structure involved:

let us write $J^r(n,p)$ for $J^r(\mathbb{R}^n, \mathbb{R}^p)_{0,0}$, and let $L^r(n) \subset J^r(n,n)$ be the group of r-jets at 0 of germs of diffeomorphism of $(\mathbb{R}^n,0)$ (with multiplication defined by $j^r h . j^r h' = j^r(h \circ h')$). Then $L^r(n) \times L^r(p)$ acts on $J^r(n,p)$ by $j^r f.(j^r h, j^r k) = j^r(k \circ f \circ h^{-1})$.

(To see that this action, and the multiplications in $L^r(n), L^r(p)$ are well-defined, we observe the general fact that the r-jet of a composite depends only on the r-jets of its constituents; for suppose that $f,f' : (X,x_0) \to (Y,y_0)$ have the same r-jet, and that $g,g' : (Y,y_0) \to (Z,z_0)$ also have the same r-jet. Then

$$(g \circ f)^* - (g' \circ f')^* = f^* \circ (g^* - g'^*) + (f^* - f'^*) \circ g'^*$$

so that, since $(g^* - g'^*)C_Z \subset m_Y^{r+1}$, $(f^* - f'^*)C_Y \subset m_X^{r+1}$, ($(g \circ f)^* - (g' \circ f')^* C_Z \subset m_X^{r+1}$, so that $g \circ f$, $g' \circ f'$ have the same r-jet).

Now a choice of local co-ordinates provides an isomorphism of $J^r(N,P)_{x_0,y_0}$ with $J^r(n,p)$; clearly different choices correspond to the action of $L^r(n) \times L^r(p)$ on $J^r(n,p)$, and so $p^{r,0} : J^r(N,P) \to N \times P$ may be viewed as a fibre bundle with fibre $J^r(n,p)$ and structure group $L^r(n) \times L^r(p)$.

(6.3) If $f : (N,x_0) \to (P,y_0)$ is a map-germ, then a germ $J^r f : (N,x_0) \to (J^r(N,P), j^r f)$ is defined by the assignment $x \longrightarrow j^r \tilde{f}_x$, where $\tilde{f}$ is a representative of f and $\tilde{f}_x$ is its germ at x. Clearly $J^r f$ is a germ of section for the projection $p_N \circ p^{r,0} : J^r(N,P) \to N$.

(6.4) <u>Lemma</u>

Let $f : (N,x_0) \to (P,y_0)$ be a map-germ, and let $z = j^r f$.

There are natural identifications (as $\mathbb{R}$-vector spaces)

a) $T(J^r(N,P)_{x_0})_z \cong \theta_f / m_N^{r+1} \theta_f$.

b) $T(J^r(N,P)_{x_0,y_0})_z \cong m_N \theta_f / m_N^{r+1} \theta_f$.

<u>Proof</u>

a) Let $\phi \in \theta_f$, and let $g : (N \times \mathbb{R}, x_0 \times 0) \to (P,y_0)$ be a map-germ such that $g(x,0) = f(x)$ and $\frac{\partial}{\partial u} \cdot g = \phi$ (as in (1.8)). Then the germ of path

$u \to j^r(g^u)$ (where $g^u : (N,x_0) \to (P,y_0)$ is defined by $g^u(x) = (x,u)$) gives an element $\bar{\phi} \in T(J^r(N,P)_{x_0})_z$; which is determined by ϕ , since $g^u(x) = f(x) + u\frac{\partial g}{\partial u}(x,0) + O(u^2)$ (with respect to any local co-ordinates).

Thus we can define $\pi^r : \theta_f \to T(J^r(N,P)_{x_0})_z$ by $\phi \to \bar{\phi}$. This is clearly $\mathbb{R}$-linear; and $\pi^r(\phi) = 0$ if and only if $j^r\phi_i = 0$ (where $\phi = \sum_{i=1}^{p} \phi_i . wf(\frac{\partial}{\partial y_i})$, where $\{y_1,\ldots,y_p\}$ are local co-ordinates in P) i.e. by (5.3), if and only if $\phi \in m_N^{r+1}\theta_f$.

So π^r induces an injection in a); but $T(J^r(N,P)_{x_0})_z$, $\theta_f/m_N^{r+1}\theta_f$ have the same $\mathbb{R}$-dimension, so this induced map is surjective also. It gives our identification.

b) This follows at once, by restricting π^r to $T(J^r(N,P)_{x_0,y_0})_z \subset T(J^r(N,P)_{x_0})_z$.

(6.5) <u>Corollary</u>

a) Let $(y_1,\ldots,y_p)$ be a system of local co-ordinates for (P,y_0) .

This induces a local trivialisation of $J^r(N,P)_{x_0}$ over (P,y_0) , and hence induces a splitting $T(J^r(N,P)_{x_0})_z \cong T(J^r(N,P)_{x_0,y_0})_z \oplus TP_{y_0}$, which, with respect to the identification of (6.4), is given by

$$\pi^r(\theta_f) \cong \pi^r(m_N\theta_f) \oplus \mathbb{R}\langle \pi^r(wf(\frac{\partial}{\partial y_1})),\ldots,\pi^r(wf(\frac{\partial}{\partial y_p}))\rangle \ .$$

b) Let $(x_1,\ldots,x_n)$ be a system of local co-ordinates at (N,x_0) .

This induces a local trivialisation of $J^r(N,P)$ over (N,x_0) , and hence induces a splitting $T(J^r(N,P))_z = T(J^r(N,P)_{x_0})_z \oplus TN_{x_0}$.

Let $f : (N,x_0) \to (P,y_0)$ be a map-germ. With respect to the splitting above, $T(J^rf)_{x_0}(\frac{\partial}{\partial x_i}\Big|_{x_0}) = \pi^r(tf(\frac{\partial}{\partial x_i})) \oplus \frac{\partial}{\partial x_i}\Big|_{x_0}$.

<u>Proof</u>

a) The splitting via local trivialisation is given by translations in P ; while the map-germs $g : (N\times\mathbb{R}, x_0\times 0) \to (P,y_0)$ defined by $g(x,u) = f(x) + (0,\ldots,u,\ldots,0)$ $(i=1,\ldots,p)$ give the action of these translations on f . Since $\frac{\partial}{\partial u}\cdot g = wf(\frac{\partial}{\partial y_i})$, the result follows.

b) This is a straightforward calculation.

Let $\phi_i : (N \times \mathbb{R}, x_0 \times 0) \to (N, x_0)$ by the translation flow given by $\phi_i(x,u) = x + (0,\ldots,u,\ldots,0)$ $(i = 1,\ldots,n)$. Then

$$T(J^r f)_{x_0}\left(\frac{\partial}{\partial x_i}\Big|_{x_0}\right) = T(J^r f)_{x_0}\left(\frac{d}{du}(\phi(x_0,u))\Big|_{u=0}\right)$$

$$= \frac{d}{du}(j^r f_{\phi(x_0,u)})\Big|_{u=0}$$

$$= \frac{d}{du}(j^r(f \circ \phi^{-1}_{(\ ,u)})_{\phi(x_0,u)})\Big|_{u=0} - \frac{d}{du}(j^r(f \circ \phi^{-1}_{(\ ,u)})_{x_0})\Big|_{u=0}$$

(by the chain rule)

The second term in the last expression is $-\pi^r(tf(-\frac{\partial}{\partial x_i}))$. Now, with respect to the chosen local co-ordinates, $f \circ \phi^{-1}_{(\ ,u)}$ has the same Taylor series at $\phi(x_0,u)$ as f has at x_0; so that $j^r f_{x_0} \to j^r(f \circ \phi^{-1}_{(\ ,u)})_{\phi(x_0,u)}$ is just translation with respect to the local trivialization by $(0,\ldots,u,\ldots,0)$. Thus indeed, with respect to the local trivialization splitting,

$$T(j^r f)_{x_0}\left(\frac{\partial}{\partial x_i}\Big|_{x_0}\right) = \left(\pi^r(tf(\frac{\partial}{\partial x_i})), \frac{\partial}{\partial x_i}\Big|_{x_0}\right).$$

(6.6) Now let us see how the notion of $\mathcal{K}$-equivalence gives rise to a group-action on J^r.

Let $\mathcal{R}$ be the group of germs of diffeomorphism of (N,x_0) (with multiplication given by composition). $\mathcal{R}$ acts on germs $f : (N,x_0) \to (P,y_0)$ by $f.h = f \circ h^{-1}$.

Let $\mathcal{C}$ be the subgroup of the group of germs of diffeomorphism of $(N \times P, x_0 \times y_0)$ whose elements H are of the form $H(x,y) = (x, H_x(y))$ with H_x a diffeomorphism-germ $(P,y_0) \to (P,y_0)$ for each $x \in (N,x_0)$. (thus, in the terminology on §4, H is a $\mathcal{C}$-equivalence). $\mathcal{C}$ acts on germs $f : (N,x_0) \to (P,y_0)$ by $(1_N, f.H) = H \circ (1_N, f)$.

Now let $\mathcal{K} = \mathcal{R}.\mathcal{C}$ (semi-direct product). This acts on germs $f : (N,x_0) \to (P,y_0)$ by $f.(h.H) = (f.h).H$, that is, by $(1_N, f.(h.H)) = H \circ (1_N, f \circ h^{-1})$. Thus, in the terminology of §4, $(h, H \circ (h \times 1_P))$ is a $\mathcal{K}$-equivalence between f and $f.(h.H)$; and similarly if (h,H) is a $\mathcal{K}$-equivalence between f and f', then $f' = f.(h.H \circ (h^{-1} \times 1_P))$.

Now let $\mathcal{R}^r \subset J^r(N,N)_{x_0,x_0}$ be the group of r-jets of elements of $\mathcal{R}$ (with multiplication defined by $j^r h.j^r h' = j^r(h \circ h')$). $\mathcal{R}^r$ acts on J^r by $j^r f.j^r h = j^r(f.h)$.

Also, let $\mathcal{C}^r \subset J^r(N \times P, N \times P)_{x_0 \times y_0, x_0 \times y_0}$ be the group of r-jets of elements of $\mathcal{C}$. $\mathcal{C}^r$ acts on J^r by $j^r f.j^r H = j^r(f.H)$.

Finally, let $\mathcal{K}^r = \mathcal{R}^r.\mathcal{C}^r$, which acts on J^r by $j^r f.(j^r h.j^r H) = j^r(f.(h.H))$.

(The multiplications in $\mathcal{R}^r, \mathcal{C}^r$, and their actions on J^r , are well-defined by the argument of (6.2)).

It is clear that (with respect to any systems of local co-ordinates for (N,x_0) , (P,y_0)) $\mathcal{R}^r, \mathcal{C}^r, \mathcal{K}^r$ are algebraic groups acting algebraically on J^r . We shall write $z\mathcal{R}^r, z\mathcal{C}^r, z\mathcal{K}^r$ for their orbits through $z \in J^r$. By a theorem of Borel [Bo], these orbits are actually submanifolds. We calculate their tangent spaces as follows:

(6.7) Lemma

If $f : (N,x_0) \to (P,y_0)$ is a map-germ, and $z = j^r f$, then

a) $T(z\mathcal{R}^r)_z = \pi^r(tf(m_N\theta_N))$

b) $T(z\mathcal{C}^r)_z = \pi^r(f^*(m_P)\theta_f)$

c) $T(z\mathcal{K}^r)_z = \pi^r(tf(m_N\theta_N) + f^*(m_P)\theta_f)$

Proof

We use the following: if G is a Lie group acting on a manifold M , then $T(xG)_x = T\alpha_x^G(TG_1)$ for any $x \in M$, where $\alpha_x^G : G \to M$ is the mapping $g \to x.g$

a) Since $\mathcal{R}^r$ is an open subset of $J^r(N,N)_{x_0,x_0}$, $T\mathcal{R}^r_{1_N} = \pi^r(m_N\theta_N)$.

Now the following diagram is clearly commutative:

$$\begin{array}{ccc} m_N\theta_N & \xrightarrow{tf} & m_N\theta_f \\ \downarrow \pi^r & & \downarrow \pi^r \\ T\mathcal{R}^r_{1_N} & \xrightarrow{T\alpha_x^{\mathcal{R}}} & TJ^r_z \end{array}$$

and so $T(z\mathcal{R}^r)_z = \pi^r(tf(m_N\theta_N))$.

b) Any element ϕ of $T\mathcal{C}^r_{1_{N\times P}}$ may be represented by a path $u \to j^r(H^u)$, where $H^u \in \mathcal{C}$, $H^0 = 1_{N\times P}$. Let $H : (N\times P\times \mathbb{R}, x_0\times y_0\times 0) \to (N\times P, x_0\times y_0)$ be defined by $H(x,y,u) = H^u(x,y)$. Then $\frac{\partial}{\partial u}.H \in m_P\theta_{N\times P, p_N}$ (where $p_N : (N\times P, x_0\times y_0) \to (N,x_0)$ is the natural projection), and $\pi^r(\frac{\partial}{\partial u}.H) = \phi$.

Thus $T\mathcal{C}^r_{1_{N\times P}} = \pi^r(m_P\theta_{N\times P, p_N})$.

Now the following diagram is commutative:

$$\begin{array}{ccc} m_P\theta_{N\times\mathbb{R},\, p_N} & \xrightarrow{\quad K \quad} & f^*(m_P)\theta_f \\ \Big\downarrow \pi^r & & \Big\downarrow \pi^r \\ T\mathcal{C}^r_{1_{N\times P}} & \xrightarrow{\quad T\alpha^{\mathcal{C}}_z \quad} & TJ^r_z \end{array}$$

(where K is defined by $\eta \to \eta\circ(1,f)$)

so, since K is clearly surjective, $T\mathcal{C}^r_{1_{N\times P}} = \pi^r(f^*(m_P)\theta_f)$.

c) Clearly $T\mathcal{K}^r_{1_N.1_{N\times P}} = T\mathcal{R}^r_{1_N} \oplus T\mathcal{C}^r_{1_{N\times P}}$; and, since

$$\alpha^{\mathcal{K}}_z(h.H) = (z.h).H\,,\ T_1\alpha^{\mathcal{K}}_z = T_1\alpha^{\mathcal{R}}_z + T_1\alpha^{\mathcal{C}}_z$$

Thus
$$\begin{aligned} T(z\mathcal{K}^r)_z &= T(z\mathcal{R}^r)_z + T(z\mathcal{C}^r)_z \\ &= \pi^r(tf(m_N\theta_N) + f^*(m_P)\theta_f)\,. \end{aligned}$$

Although the next is rather a side-issue as far as this book is concerned, it seems worthwhile to include it for completeness:

(6.8) Lemma

If f is $\mathcal{K}$-determined by its r-jet, then

$$m_N^{r+1}\theta_f \subset tf(m_N\theta_N) + f^*(m_P)\theta_f\,.$$

Proof

If $f : (N,x_0) \to (P,y_0)$ is $\mathcal{K}$-determined by its r-jet, let $z = j^{r+1}f$, and let $E = (p^{r+1,r})^{-1}p^{r+1,r}(z)$, so E is the set of (r+1)-jets of all germs with the same r-jet as f . Since f is $\mathcal{K}$-determined by $j^r f$, $E \subset z\mathcal{K}^{r+1}$; so that $TE_z \subset T(z\mathcal{K}^{r+1})_z$. But clearly $TE_z = \pi^{r+1}(m_N^{r+1}\theta_f)$, so that by (6.7) c) we have

$$\pi^{r+1}(m_N^{r+1}\theta_f) \subset \pi^{r+1}(tf(m_N\theta_N) + f^*(m_P)\theta_f)$$

from which we deduce

$$m_N^{r+1}\theta_f \subset tf(m_N\theta_N) + (f^*m_P + m_N^{r+2}).\theta_f \ .$$

By (1.10), then, the C_N-module

$$\{tf(m_N\theta_N) + (f^*(m_P) + m_N^{r+1})\theta_f\}/\{tf(m_N\theta_N) + f^*(m_P)\theta_f\}$$

is zero; so that $m_N^{r+1}\theta_f \subset tf(m_N\theta_N) + f^*(m_P)\theta_f$, as required.

(6.9) Corollary

A map-germ $f : (N,x_0) \to (P,y_0)$ is $\mathcal{K}$-determined by a finite jet if and only if it is of FST .

Proof

'If' was proved in (5.7).

'Only if' ; if f is $\mathcal{K}$-determined by its r-jet, then by (6.8) $m_N^{r+1}\theta_f \subset tf(m_N\theta_N) + f^*(m_P)\theta_f$, so that certainly $\dim_{\mathbb{R}} \{\theta_f/tf(\theta_N) + f^*(m_P)\theta_f\} < \infty$; i.e. f is of FST .

(6.10) The group $\mathcal{K}$ is a subgroup of the pseudogroup $\mathcal{K}_{N,P} = \mathcal{R}_{N,N}.\mathcal{C}_{N\times P,N\times P} = \{h.H | p_N \text{ source } (H) = \text{source } (h)\}$, where $\mathcal{R}_{N,N}$ is the pseudogroup of point-germs of diffeomorphisms $N \to N$ and $\mathcal{C}_{N\times P,N\times P}$ is the pseudogroup of germs H of diffeomorphism $(N \times P, x' \times y') \to (N \times P, x' \times y'')$, for any $x' \in N$, $y', y'' \in P$, of the form $H(x,y) = (x, H_x(y))$ with H_x a diffeomorphism-germ $(P,y') \to (P,y'')$ for all $x \in (N,x')$.

$\mathcal{K}_{N,P}$ acts on point-germs $f : N \to P$ by $(1_N, f.(h.H)) = H \circ (1_N, f \circ h^{-1})$, whenever this composition is defined.

Also of interest is the subpseudogroup

$$\mathcal{K}_{x_0,P} = \{h.H \in \mathcal{K}_{N,P} \mid \text{source } (h) = \text{target } (h) = x_0\}$$

which acts on germs $f : (N,x_0) \to P$.

Taking r-jets, we obtain pseudogroup actions $\mathcal{K}^r_{N,P}$ on $J^r(N,P)$, $\mathcal{K}^r_{x_0,P}$ on $J^r(N,P)_{x_0}$. Let $z\mathcal{K}^r_{N,P} \subset J^r(N,P)$ be the equivalence class of $z \in J^r(N,P)$ under the action of $\mathcal{K}^r_{N,P}$, and let $z\mathcal{K}^r_{x_0,P}$ be the equivalence class of $z \in J^r(N,P)_{x_0}$ under the action of $\mathcal{K}^r_{x_0,P}$.

Choosing local co-ordinates, we see at once that, with respect to the trivialization $J^r(N,P)|U \times V = U \times V \times J^r$, $z\mathcal{K}^r_{x_0,P}|\{x_0\} \times V = \{x_0\} \times V \times z\mathcal{K}^r$, $z\mathcal{K}^r_{N,P}|U \times V = U \times V \times z\mathcal{K}^r$ (for the action of an element of $(\mathcal{K}_{N,P})_{x,y}$ for x near x_0 , y near y_0 , may always be expressed as the composite of the action of an element of $\mathcal{K}$ with translations (with respect to the local co-ordinates) $x \to x_0$, $y_0 \to y$). Thus $z\mathcal{K}^r_{x_0,P}$, $z\mathcal{K}^r_{N,P}$ are submanifolds of $J^r(N,P)_{x_0}$, $J^r(N,P)$ respectively. It follows from (6.5) a) and (6.7) that

$$T(z\mathcal{K}^r_{x_0,P})_z = \pi^r(tf(m_N\theta_N) + wf(\theta_P) + (f^*m_P)\theta_f) .$$

At last we can state and prove our transversality theorem:

(6.11) <u>Theorem</u>

Let $f : (N,x_0) \to (P,y_0)$ be a map-germ of FST .

Let $r \geq \chi_f + 1$.

Let $F : (N \times U, x_0 \times u_0) \to (P \times U, y_0 \times u_0)$ be defined by $F(x,u) = (x,g(x,u))$, where $g : (N \times U, x_0 \times u_0) \to (P,y_0)$ is a map-germ such that $g(x,u_0) = f(x)$.

Let the germ $J^r_*F : (N \times U, x_0 \times u_0) \to (J^r(N,P), j^rf)$ be defined by $J^r_*F(x,u) = j^r\tilde{F}^u_x$, where $\tilde{F}$ is a representative of F , $\tilde{F}^u$ is defined by $\tilde{F}^u(x) = \tilde{F}(x,u)$, and $\tilde{F}^u_x$ is the germ of $\tilde{F}^u$ at x .

Then F is stable if and only if J^r_*F is transverse to $(j^rf)\mathcal{K}^r_{N,P}$ at j^rf .

<u>Proof</u>

It is clear that J^r_*F is the composite

$$(N \times U, x_0 \times u_0) \xrightarrow{J^r g} J^r(N \times U, P) \xrightarrow{a} J^r(N,P) \times U \xrightarrow{b} J^r(N,P)$$

where a is 'forgetting derivatives in the U-directions' and b is projection. So, if $\{x_1,\ldots,x_n\}$ is a system of local co-ordinates at (N,x_0) , $\{u_1,\ldots,u_k\}$ is a system of local co-ordinates at (U,u_0) , then it follows easily from our calculation (6.5) b) (applied to g) that, with repsect to the splittings induced by the co-ordinates,

$$\begin{cases} T(J^r_*F)_{x_0\times u_0}(\frac{\partial}{\partial x_i}|_{x_0}) = (\pi^r(tf(\frac{\partial}{\partial x_i})), \frac{\partial}{\partial x_i}|_{x_0}) \\ T(J^r_*F)_{x_0\times u_0}(\frac{\partial}{\partial u_i}|_{u_0}) = (\pi^r \frac{\partial}{\partial u_i}\cdot g, 0) \end{cases}$$

so that

$$\operatorname{Im} T(J^r_*F)_{x_0\times u_0} = \mathbb{R}\langle \pi^r(tf(\frac{\partial}{\partial x_i})), \pi^r(\frac{\partial}{\partial u_i}\cdot g)\rangle \oplus TN_{x_0} .$$

Thus J^r_*F is transverse to $z\mathcal{K}^r_{N,P}$ if and only if

$$\mathbb{R}\langle\pi^r(tf(\frac{\partial}{\partial x_i})), \pi^r(\frac{\partial}{\partial u_i}\cdot g)\rangle + T(z\mathcal{K}^r_{x_0,P}) = T(J^r(N,P)_{x_0})_z$$

But, by our formula for $T(z\mathcal{K}^r_{x_0,P})_z$, and since $\operatorname{Ker} \pi^r = m_N^{r+1}\theta_f$, this is equivalent to

$$\theta_f = tf(\theta_N) + wf(\theta_P) + \mathbb{R}\langle\frac{\partial}{\partial u_i}\cdot g\rangle + (f^*(m_P) + m_N^{r+1})\theta_f ,$$

which in turn is equivalent to

$$\theta_f = tf(\theta_N) + wf(\theta_P) + \mathbb{R}\langle\frac{\partial}{\partial u_i}\cdot g\rangle + f^*(m_P)\theta_f ,$$

since $m_N^{r+1}\theta_f \subset tf(\theta_N) + f^*(m_P)\theta_f$ from our hypothesis and (1.12).

But the last equation is equivalent to the surjectivity of

$$\rho_{F,f} : T(P\times U)_{y_0\times u_0} \to \theta_f/tf(\theta_N) + f^*(m_P)\theta_f$$

since, if $y_1,\ldots,y_p$ is a system of local co-ordinates at (P,y_0) ,

$$\rho_{F,f}(\frac{\partial}{\partial y_i}|_{y_0}) = [wf(\frac{\partial}{\partial y_i})] \text{ and } \rho_{F,f}(\frac{\partial}{\partial u_j}|_{u_0}) = [-\frac{\partial}{\partial u_j}\cdot g] .$$

By (2.6), this is equivalent to F being stable, and so the theorem is proved.

(6.12) Remark

Thus F is stable if and only if J^r_*F acts as a 'slice' to the $\mathcal{K}^r_{N,P}$-equivalence classes; so we expect all 'nearby' classes to be represented by $\tilde{F}^u_x$ for some $(x,u) \in N\times U$ 'near' (x_0,u_0) .

We take this as justification for our remarks about 'contact-versality' in §4.

§7. Genericity

In this section we show that map-germs of FST are 'generic'. More precisely, we will show that the map-germs not of FST form a subset of infinite codimension, in a certain sense, of the space of map-germs (and this even with respect to the 'obvious' topology - comparing all partial derivatives at the source - which we dismissed in §0 as apparently too coarse!).

(7.1) Let $z \in J^r(n,p)$, and let $f : (\mathbb{R}^n,0) \to (\mathbb{R}^p,0)$ be a map-germ such that $j^r f = z$.

Define $\chi_z = \dim_{\mathbb{R}}\{\theta_f/tf(\theta_N) + (f^*m_P + m_N^r)\theta_f\}$ (where we write $N = \mathbb{R}^n$).

To see that this is well-defined, suppose that $j^r f = j^r f'$. Then, by (5.4),

$$tf'(\theta_N) + (f'^*m_P + m_N^r)\theta_f \equiv tf(\theta_N) + (f^*m_P + m_N^r)\theta_f ,$$

so that

$$\dim_{\mathbb{R}}\{\theta_f/tf(\theta_N) + (f^*m_P + m_N^r)\theta_f\} = \dim_{\mathbb{R}}\{\theta_{f'}/tf'(\theta_N) + (f'^*m_P + m_N^r)\theta_{f'}\} .$$

Define $W^r(n,p) = \{z \in J^r(n,p) \mid \chi_z \geq r\}$.

We observe the following facts about $W^r(n,p)$:

a) If $z \notin W^r(n,p)$, then z is $\mathcal{K}$-determining (that is, if f is such that $j^r f = z$, then f is $\mathcal{K}$-determined by its r-jet.

This follows since, if $\chi_z \leq r-1$, then, by (1.12),

$$m_N^{r-1}\theta_f \subset tf(\theta_N) + (f^*m_P)\theta_f ,$$

whence, by (5.6), f is indeed $\mathcal{K}$-determined by its r-jet).

b) $W^r(n,p)$ is $\mathcal{K}^r$-invariant.

(For if $z = j^r f$, then, by (1.12), $\chi_z \geq r$ if and only if $\dim_{\mathbb{R}}\{\theta_f/tf(\theta_N) + (f^*m_P + m_N^{r+1})\theta_f\} \geq r$. But, by (6.7) (and (6.4)) this number is the codimension of $z\mathcal{K}^r$, and so is preserved by $\mathcal{K}^r$).

c) $W^r(n,p)$ is a homogeneous real-algebraic variety in $J^r(n,p)$.

(For $W^r(n,p)$ is defined by the condition that a linear map with polynomial coefficients has corank at least r).

(7.2) Our 'genericity' theorem is:

Theorem

$$\lim_{r\to\infty} \text{codim. } W^r(n,p) = \infty .$$

Proof

Let us write, for convenience, $W^r = W^r(n,p)$.

Let $r \geq s$. Then we have, clearly, that $(p^{r,s})^{-1}W^s \supset W^r$, so that codim. W^r is non-decreasing with r . It will therefore be sufficient to show that for each r there exists $r' > r$ such that codim. $W^{r'} >$ codim. W^r .

To do this, it will be enough to show that for each irreducible component C of W^r there exists $r_c > r$ such that $(p^{r_c,r})^{-1}C - W^{r_c} \neq \phi$; for then $(p^{r_c,r})^{-1}C \cap W^{r_c}$ is a union of irreducible components of W^{r_c} , each of which, by the above, is of smaller dimension than dim. $(p^{r_c,r})^{-1}C$, that is, of greater codimension than codim. C . Thus, if $r' = \max_c r_c$, each irreducible component of $W^{r'}$ is of greater codimension than some irreducible component of W^r , so that indeed codim. $W^{r'} >$ codim. W^r .

Finally, then, it will clearly be enough to find $r' > r$ such that for any $z \in W^r$ there exists $z' \in (p^{r',r})^{-1}z - W^{r'}$.

We proceed as follows :

let $f : (\mathbb{R}^n,0) \to (\mathbb{R}^p,0)$ be the germ defined by

$$f(x_1,\ldots,x_n) = \begin{cases} (x_1^{r+1},\ldots,x_n^{r+1},0,\ldots,0) & \text{if } p \geq n \\ (x_1^{r+1}+\ldots+x_{n-p+1}^{r+1},\ldots,x_i^{r+1}+\ldots+x_{n-p+i}^{r+1},\ldots,x_p^{r+1}+\ldots+x_n^{r+1}) & \text{if } p \leq n. \end{cases}$$

In the case $p \geq n$, $f^*(y_i) = x_i^{r+1}$ for $i \leq n$, and so, since $m_N^{n(r+1)}$ is generated by polynomials of the form $\prod_{i=1}^{n} x_i^{p_i}$, where $\sum_{i=1}^{n} p_i = n(r+1)$, it follows that $m_N^{n(r+1)} \subset f^*m_P$. Thus $m_N^{n(r+1)}\theta_f \subset (f^*m_P)\theta_f$; in particular $\chi_f < \infty$.

In the case $p \leq n$: we claim that

$$x_i^{r+1}wf(\frac{\partial}{\partial y_j}) \subset tf(\theta_N) + (f^*m_P)\theta_f \qquad (i=1,\ldots,n\,;\, j=1,\ldots,p) .$$

By symmetry, it will be enough to show this in the case $j = 1$.

Let us write $y'_i = y_i - y_{i-1}$ for $i=2,\dots,p$. With respect to the co-ordinates $(y_1, y'_2, \dots, y'_p)$, f has the form

$$f(x_1,\dots,x_n) = (x_1^{r+1} + \dots + x_{k+1}^{r+1}, \dots, x_{k+i+1}^{r+1} - x_i^{r+1}, \dots, x_n^{r+1} - x_p^{r+1})$$

(where we write $k = n - p$).

First consider the case $1 \le i \le k+1$.

Let q be such that $(q-1)(k+1) + i < p$, $q(k+1) + i \ge p$.

We have

$$\text{(o)} \qquad \frac{1}{r+1} \cdot x_i tf(\frac{\partial}{\partial x_i}) = x_i^{r+1}\{wf(\frac{\partial}{\partial y_1}) - wf(\frac{\partial}{\partial y'_{i+1}})$$

and, for $1 \le s \le q-1$,

$$\text{(s)} \quad \frac{1}{r+1} \cdot x_{s(k+1)+i} tf(\frac{\partial}{\partial x_{s(k+1)+i}}) = x_{s(k+1)+i}^{r+1}\{wf(\frac{\partial}{\partial y'_{(s-1)(k+1)+i+1}}) - wf(\frac{\partial}{\partial y'_{s(k+1)+i+1}})\}$$

and, finally,

$$\text{(q)} \qquad \frac{1}{r+1} \cdot x_{q(k+1)+i} tf(\frac{\partial}{\partial x_{q(k+1)+i}}) = x_{q(k+1)+i}^{r+1}\{wf(\frac{\partial}{\partial y'_{(q-1)(k+1)+i+1}})\}$$

Adding these $q+1$ equations, we obtain

$$x_i^{r+1} wf(\frac{\partial}{\partial y_1}) = \sum_{s=0}^{q} \frac{x_{s(k+1)+i}}{r+1} tf(\frac{\partial}{\partial x_{s(k+1)+i}}) - \sum_{s=0}^{q-1} (x_{(s+1)(k+1)+i}^{r+1} - x_{s(k+1)+i}^{r+1}) wf(\frac{\partial}{\partial y'_{s(k+1)+i+1}})$$

Now the first sum on the LHS is an element of $tf(\theta_N)$; while, since $x_{(s+1)(k+1)+i}^{r+1} - x_{s(k+1)+i}^{r+1} = f^*(y'_{s(k+1)+i+1})$ for $q-1 \ge s \ge 0$, the second sum on the LHS is an element of $(f^*m_p)\theta_f$. So we have shown

$$x_i^{r+1} wf(\frac{\partial}{\partial y_1}) \subset tf(\theta_N) + (f^*m_p)\theta_f \quad \text{for} \quad 1 \le i \le k+1 .$$

Now consider the case $i > k+1$; we may write $i = q(k+1) + j$, where $q > 0$ and $1 \le j \le k+1$.

Then

$$x_i^{r+1} wf(\frac{\partial}{\partial y_1}) = \sum_{s=0}^{q-1} (x_{(s+1)(k+1)+j}^{r+1} - x_{s(k+1)+j}^{r+1}) wf(\frac{\partial}{\partial y_1}) + x_j^{r+1} wf(\frac{\partial}{\partial y_1}) .$$

But $x_{(s+1)(k+1)+j}^{r+1} - x_{s(k+1)+j}^{r+1} = f^*(y'_{s(k+1)+j+1})$, and $x_j^{r+1} wf(\frac{\partial}{\partial y_1}) \in tf(\theta_N) + (f^*m_p)\theta_f$, by the previous analysis, so that

$x_i^{r+1} wf(\frac{\partial}{\partial y_1}) \in tf(\theta_N) + (f^*m_P)\theta_f$ also, and thus the claim is proved.

It follows easily that $m_N^{n(r+1)}\theta_f \subset tf(\theta_N) + (f^*m_P)\theta_f$, so that in particular $\chi_f < \infty$.

In either case, then, let $r' = \chi_f + 1$, and let $z_0 = j^{r'}f$; so that $z_0 \notin W^{r'}$ (by (5.6) and (7.1) a)).

Let z be any element of W^r , and let $\tilde{z} \in (p^{r',r})^{-1}z$. The line L through $\tilde{z}$ and z_0 is not contained in $W^{r'}$, since $z_0 \notin W^{r'}$. Hence, since $W^{r'}$ is algebraic, $L \cap W^{r'}$ consists of at most finitely many points. Certainly, then, there exists $\lambda \in (0,1]$ such that $z' = \frac{1-\lambda}{\lambda} . z_0 + \tilde{z}$ is not contained in $W^{r'}$. But $p^{r',r}(z') = z$ (since $p^{r',r}(z_0) = j^r f = 0$). Thus we have constructed the required $r' > r$, and the proof is complete.

CHAPTER IV

Proof of the Topological Stability Theorem

by

Eduard Looijenga

Throughout this chapter N and P are fixed manifolds and their dimensions are denoted by n and p respectively.

§1 Multi-transversality

In the next sections we will encounter the following situation: a stratified subset $(A, \mathcal{A})$ of the jet space and a smooth mapping $f : N \to P$ such that the jet section $J^{\ell}f : N \to J^{\ell}(N, P)$ is transversal to the strata of $\mathcal{A}$. Following (I, 1.4), $(A, \mathcal{A})$ then pulls back via $J^{\ell}f$ to a stratified subset $(B, \mathcal{B})$ of N. It would be nice if we could refine $\mathcal{B}$ to a stratification $\mathcal{B}'$ of B with the property that the f-images of any two members of $\mathcal{B}'$ are either disjoint or equal. This is possible if the f-images of the strata of $\mathcal{B}$ intersect in P in a 'regular' way. In this section we will show that under fairly mild conditions this last 'regular' intersection property holds if we impose a certain transversality condition on f. This transversality condition (usually called multi-transversality) was introduced by Mather, who also showed that it is a 'generic' property. This corresponds to our theorem (1.1), whose proof is rather close to his. It is perhaps good to emphasize here that multi-transversality is not necessarily an open condition (even when the manifolds in question are compact), although in the last section we will meet a situation where this happens to be the case. Another main result of this section, proposition (1.5), is also concerned with this: it asserts that a stable map-germ satisfies all 'natural' multi-transversality conditions.

We begin to give the term 'regular intersection' a precise meaning. Recall that a family $(V_i)_{i\in I}$ of subspaces of a vector space V is said to be in <u>general</u> (or <u>regular</u>) <u>position</u> if the canonical map

$$V \to \bigoplus_{i\in I} V/V_i$$

is surjective, or equivalently, if $\operatorname{codim}(\cap_{i\in I} V_i) = \Sigma_{i\in I} \operatorname{codim} V_i$. Clearly, this condition implies that there are at most dim V subspaces V_i not equal to V. This notion has its counterpart in the context of stratified sets.

Suppose we are given a smooth mapping $f : N \to P$ and a stratified subset $(B, \mathcal{B})$ of N. Denote for any $x \in B$ the stratum of $\mathcal{B}$ which contains x by X_x. Then $(B, \mathcal{B})$ has regular intersections relative f, if for any $y \in f(B)$ and any subset I of $f^{-1}(y)$ the subspaces $\{Tf(T_x X_x)\}_{x \in I}$ of $T_y P$ are in general position. By the remark above it suffices to verify this condition for subsets I of cardinal $\leq p+1$. In particular, each fibre of $f|B$ then contains at most p points where f is of rank $< p$. Assume moreover that in this situation any $y \in P$ possesses a neighbourhood such that its f-counterimage intersects only finitely many strata. This is for instance the case when $f|B$ is proper. Then there exists a refinement $\mathcal{B}_f$ of $\mathcal{B}$ with the following two properties: (i) any two strata of $\mathcal{B}_f$ have either equal or disjoint images under f and (ii) any refinement of $\mathcal{B}$ having property (i) refines $\mathcal{B}_f$. More explicitly, $\mathcal{B}_f$ may be characterized by the property that a member of $\mathcal{B}_f$ is an indecomposable element of the Boolean algebra generated by $\{X \cap f^{-1}f(Y) : X, Y \in \mathcal{B}\}$. Hence a member of $\mathcal{B}_f$ is locally of the form $f^{-1}(f(X_1) \cap \ldots \cap f(X_k))$ with $X_1, \ldots, X_k \in \mathcal{B}$. The regular intersection property implies that the sets of this form are smooth. Clearly, $\mathcal{B}_f$ is locally finite, so we conclude that $\mathcal{B}_f$ is a stratification of B. We note in passing the naturality of this construction with respect to subsets V in P:

$$(\mathcal{B}|f^{-1}V)_{f|f^{-1}V} = \mathcal{B}_f|f^{-1}V .$$

$\mathcal{B}_f$ need not satisfy the frontier property, even if we replace strata by their connected components, as is illustrated below.

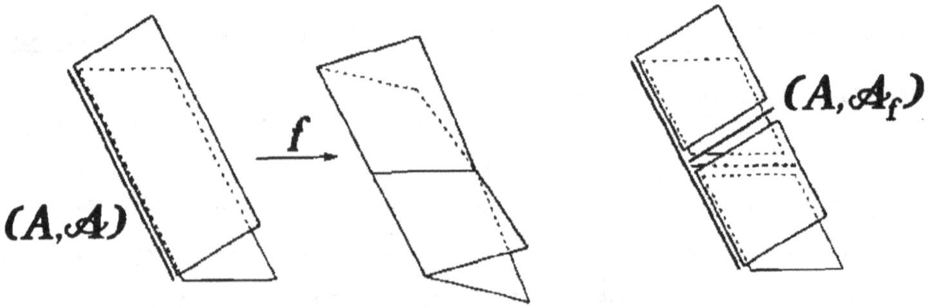

As mentioned earlier, the stratified subset $(B, \mathcal{B})$ of N will usually be obtained as the transversal counterimage of a stratified subset of the jet bundle. This suggests the following definition.

Let $(A, \mathcal{A})$ be a stratified subset of $J^\ell(N, P)$. A smooth mapping $f : N \to P$ is called <u>multi-transverse with respect to</u> $(A, \mathcal{A})$ if $J^\ell f$ is transversal to the strata of $\mathcal{A}$ and moreover, the (hence defined) stratified subset $(J^\ell f)^{-1}(A, \mathcal{A})$ of N has regular intersections relative f.

Our first aim is to show that if $(A, \mathcal{A})$ satisfies a certain condition, 'most' mappings will be multi-transverse with respect to $(A, \mathcal{A})$. To make this precise, we need a topology on the function space $C^\infty(N, P)$. For our purposes the Whitney topology is best. The subsets

$$M(\Omega) = \{f \in C^\infty(N, P) : J^\ell f(N) \subseteq \Omega\},$$

($\ell = 0, 1, 2, \ldots$, Ω an open subset of $J^\ell(N, P)$) form a basis for this topology. Following [Morlet] $C^\infty(N, P)$ with the Whitney topology is a Baire space, i.e. a countable intersection of open - dense subsets of $C^\infty(N, P)$ is dense. This leads to the following definition.

Let U be an open subset of $C^\infty(N, P)$. A subset of U is called <u>generic</u> in U if it contains a countable intersection of open - dense subsets of U. Clearly, a countable interesection of such sets is again generic.

(1.1) Theorem. <u>Let</u> $(A, \mathcal{A})$ <u>be a stratified subset of</u> $J^\ell(N, P)$ <u>with the property that for any stratum</u> $X \in \mathcal{A}$, <u>the natural projection</u> $\pi_P | X : X \to P$ <u>is a submersion. Then the set of</u> $f \in C^\infty(N, P)$ <u>which are multi-transverse with respect to</u> $(A, \mathcal{A})$ <u>is generic in</u> $C^\infty(N, P)$.

We prove (1.1) via two lemmas. In the first one we deal with an auxiliary stratification $\mathcal{C}_r$ of the r - fold Cartesian product $A^r \subset J^\ell(N, P)^r$ which is constructed as follows. Let $\mathcal{D}_r$ be the 'diagonal stratification' of P^r : two points of P^r belong to the same stratum of $\mathcal{D}_r$ if and only if they have the same orbit type with respect to the natural action of the symmetric group on N^r.

It is not difficult to verify that this defines a stratification (see also the discussion at the end of I, 2)). Since $\pi_P : J^\ell(N, P) \to P$ maps the strata of $\mathcal{A}$ submersively to P, $\pi_P^r : J^\ell(N, P)^r \to P^r$ maps the strata of $\mathcal{A}^r$ submersively to P^r. Hence A^r and $(\pi_P^r)^{-1}\mathfrak{D}_r$ are in regular position so that their intersection, which we denote by $\mathcal{E}_r$, is a stratification of A^r.

We further put $N^{(r)} = \{(x_1, \dots, x_r) \in N^r : i \neq j \Rightarrow x_i \neq x_j\}$, and for any $f \in C^\infty(N, P)$ we let ${}_rJ^\ell f : N^{(r)} \to J^\ell(N, P)^r$ denote the restriction of $(J^\ell f)^r$.

(1.2) Lemma. <u>Let</u> $f \in C^\infty(N, P)$ <u>and suppose that</u> ${}_rJ^\ell f$ <u>is transverse to</u> $(A^r, \mathcal{E}_r)$ <u>for</u> $r = 1, \dots, p+1$. <u>Then</u> f <u>is multi-transverse with respect to</u> $(A, \mathcal{A})$.

Proof. Since ${}_rJ^\ell f$ is transverse to $(A^r, \mathcal{E}_r)$ $(r = 1, \dots, p+1)$, it is in particular transverse to $(A^r, \mathcal{A}^r)$, and the restriction of $({}_rJ^\ell f)^{-1}(A^r, \mathcal{A}^r)$ is transverse to $(\pi_P^r)^{-1}\mathfrak{D}_r$. The latter is equivalent to the condition that the restriction of $\pi_P^r \circ {}_rJ^\ell f$ to the strata of $({}_rJ^\ell f)^{-1}(A^r, \mathcal{A}^r)$ is transverse to $\mathfrak{D}_r$. Since $\pi_P^r \circ {}_rJ^\ell f$ is simply the restriction of $f^r : N^r \to P^r$ to $N^{(r)}$ it is easily seen that this means that $(J^\ell f)^{-1}(A, \mathcal{A})$ has regular intersections relative f at subsets of N of cardinal $r \leq p+1$. In view of an earlier remark this proves the lemma.

(1.3) Lemma. <u>Let</u> M <u>be a submanifold of</u> $J^\ell(N, P)^r$, <u>let</u> $z \in M$ <u>and</u> $(x_1^o, \dots, x_r^o) \in N^{(r)}$. <u>There exist neighbourhoods</u> U <u>of</u> $(x_1^o, \dots, x_r^o)$ <u>in</u> $N^{(r)}$ <u>and</u> V <u>of</u> z <u>in</u> M <u>such that the set</u> $\Omega(U, V)$ <u>of</u> $f \in C^\infty(N, P)$ <u>with</u> ${}_rJ^\ell f|U$ <u>transverse to</u> M <u>on</u> V <u>is open and dense in</u> $C^\infty(N, P)$.

Proof. Choose a compact neighbourhood U_p of x_p^o in a coordinate patch for N such that $U_1, \dots, U_r$ are disjoint, and put $U = U_1 \times \dots \times U_r \subset N^{(r)}$. To prove that $\Omega(U, M)$ is dense in $C^\infty(N, P)$, pick an $f \in C^\infty(N, P)$. It is not difficult to construct a k-parameter family $(f_t \in C^\infty(N, P))_{t \in \mathbb{R}^k}$ with $k = \dim J^\ell(N, P)^r$ such that (i) $f_o = f$, (ii) $(x, t) \mapsto f_t(x)$ is smooth, (iii) f and f_t agree outside some compact subset of N for all t, and (iv) $(x_1, \ldots, x_r, t) \xrightarrow{J} {}_rJ^\ell f_t(x_1, \ldots, x_r) \in J^\ell(N, P)^r$ is a submersion on a neighbourhood W of $U \times \{0\}$. (This can be done by means of the coordinates on $U_1, \ldots, U_r$ and finitely many coordinate patches for P covering the compact set $f(U_1) \cup \ldots \cup f(U_r)$ which so determine a finite number of coordinate patches for the jet space $J^\ell(N, P)$. The reader who wishes to see a precise proof is referred to [Mather 5].) Then $W \cap J^{-1}M$ is a submanifold of W. Now, if t is a regular value of the composite $W \cap J^{-1}M \subset N^{(r)} \times \mathbb{R}^k \to \mathbb{R}^k$ then $W \cap J^{-1}M$ is transverse to $N^{(r)} \times \{t\}$ and hence ${}_rJ^\ell f_t|U$ is transverse to M. Since by Sard's theorem such regular values are dense in $\mathbb{R}^k$, it follows that f lies in the closure of $\Omega^\ell(U, M)$. This proves that $\Omega(U, M)$ is dense in $C^\infty(N, P)$. In particular, for any subset V of M, $\Omega(U, V)$ is dense in $C^\infty(N, P)$. But if V is compact, then $\Omega(U, V)$ is also open.

Proof of (1.1). By Lemma (1.2) it is sufficient to prove that for a submanifold M of $J^\ell(N, P)^r$ the set of $f \in C^\infty(N, P)$ with ${}_rJ^\ell f$ transverse to M is generic in $C^\infty(N, P)$. In view of lemma (1.3) and the fact that $N^{(r)} \times M$ has a countable basis we can cover $N^{(r)} \times M$ by countably many set $\{U_i \times V_i\}_{i=1}^\infty$ with $\Omega(U_i, V_i)$ open and dense in $C^\infty(N, P)$. Clearly, $\Omega(N^{(r)}, M) = \bigcap_{i=1}^\infty \Omega(U_i, V_i)$, proving that $\Omega(N^{(r)}, M)$ is generic in $C^\infty(N, P)$.

As we have now shown that multi-transversality is a generic property, we may expect that any stable map-germ has a representative satisfying any finite number of natural multi-transversality conditions. This is made precise by the following

(1.4) Proposition. <u>Let</u> $(A, \mathcal{A})$ <u>be a stratified subset of</u> $J^\ell(N, P)$ <u>which is invariant under the natural action of</u> diff $N \times$ diff P <u>on</u> $J^\ell(N, P)$. <u>Then any stable map-germ</u> $f : (N, x) \to (P, y)$ <u>has a representative which is multi-transverse with respect to</u> $(A, \mathcal{A})$.

We remark here that by results of Mather a stronger statement is true. Namely that f admits a representative which is multi-transverse with respect to the (possibly uncountably many) diff $N \times$ diff P orbits in $J^\ell(N, P)$.

Proof of (1.4). Put $\ell' = \max(\ell, p + 1)$. Following (III, 5.8) $j^{\ell'}f$ is sufficient and according to (III, 6.11), $J^{\ell'}f : (N, x) \to J^{\ell'}(N, P)$ intersects the diff N × diff P orbit of $j^{\ell'}f$ transversally. Now choose a representative $\tilde{f} : U \to P$ of f. Since the hypotheses of (1.1) are satisfied we can approximate $\tilde{f}$ by a smooth mapping $\tilde{g} : U \to P$ which is multi-transverse with respect to $(A, \mathcal{A})$. By the transversality property of $J^{\ell'}f$ we may moreover require that there exists an $x' \in U$ such that $J^{\ell'}\tilde{g}(x')$ lies in the diff N × diff P orbit of $j^{\ell'}f$. Put $g = \tilde{g}_{x'}$. Since $j^{\ell'}f$ is sufficient, the fact that $j^{\ell'}f$ and $j^{\ell'}g$ lie in the same orbit implies that f is smoothly equivalent to g. The required multi-transversality property now follows from the fact that g possesses this property and the diff N × diff P-invariance of $(A, \mathcal{A})$.

(1.5) Corollary. <u>Let</u> $f : (N, x) \to (P, y)$ <u>be a map-germ of finite singularity type. Then</u> f <u>has a representative</u> $\tilde{f}$ <u>such that the restriction of</u> $\tilde{f}$ <u>to its critical set has finite fibres.</u>

Proof. According to (III, 2.8) such a map-germ possesses a stable unfolding. Of course, it suffices to prove (1.6) for this unfolding, so we may as well suppose that f is a stable map-germ. Following (1.4), f then admits a representative $\tilde{f} : U \to P$ which is multi-transverse with respect to the whole manifold $J^0(N, P) = N \times P$. In other words U has regular intersections relative $\tilde{f}$. As observed at the beginning of this section, this implies that each fibre of $\tilde{f}$ contains at most p points where $\tilde{f}$ is of rank $< p$.

We close this section with a simple, but useful lemma concerning transversality.

(1.6) Lemma. <u>Let</u> M <u>and</u> Q <u>be manifolds</u>, $F : M \to Q$ <u>a smooth mapping</u>, $R \subset Q$ <u>a submanifold and</u> $(B, \mathcal{B})$ <u>a stratified subset of</u> M <u>such that</u> $\mathcal{B}_F$ <u>is a stratification, and the restriction of</u> F <u>to any stratum of</u> $\mathcal{B}$ <u>is transverse to</u> R. <u>Then the following two conditions are equivalent.</u>

(i) Near $F^{-1}R$, $(B, \mathcal{B})$ has regular intersections relative F and $F^{-1}R$ is transverse to $\mathcal{B}_F$.

(ii) $(B, \mathcal{B})|F^{-1}R$ has regular intersections relative $F|F^{-1}R$.

(Note that our hypotheses imply that $F^{-1}R$ is a smooth submanifold of M near $B \cap F^{-1}R$, so that these conditions make sense.)

Proof. Let $y \in R$ and let I be a subset of $B \cap F^{-1}R$. As before, let, for any $x \in I$, X_x denote the stratum of $\mathcal{B}$ which contains x. Then (i) corresponds to the condition that the canonical map

$$T_y Q \rightarrow \bigoplus_{x \in I} T_y Q/TF(T_x X_x)$$

is surjective and that its kernel is transverse to $T_y R$. Likewise (ii) corresponds to the condition that

$$T_y R \rightarrow \bigoplus_{x \in I} T_y R/(TF(T_x X) \cap T_y R)$$

is surjective. Now we have canonical isomorphisms

$$T_y R/(TF(T_x X_x) \cap T_y R) \cong (T_y R + TF(T_x X_x))/TF(T_x X_x)$$
$$= T_y Q/TF(T_x X_x)$$

(since $F|X_x$ is transversal to R), so the two conditions are clearly equivalent.

§2 A stratification of the jet space

First a word of motivation. Our aim is to construct a stratification $\mathcal{A}^\ell(N, P)$, of a big open subset of $J^\ell(N, P)$, with the following rather strong property: if ℓ is sufficiently large, then, for any mapping $f : N \rightarrow P$ which is multitransverse with respect to $\mathcal{A}^\ell(N, P)$, the locally finite manifold partition $\mathcal{B} = ((J^\ell f)^{-1} \mathcal{A}^\ell(N, P))_f$ is actually a Whitney stratification which extends to a Thom stratification $(\mathcal{B}, \mathcal{B}')$ of f in the sense of (I, 3), in particular, the partition $\mathcal{B}' = \{f(X) : X \in \mathcal{B}\}$ is a Whitney stratification. Once we have established this in section 3, the topological ability theorem will be close at hand. In this section, first the construction of $A^{\ell}(N, P)$ is carried out. Unfortunately, the nature of this construction is such that it is not even obvious that $A^{\ell}(N, P)$ is a manifold partition. The proof that this is so, is also postponed to section 4. Here we settle a partial result, which is however quite crucial: if $f : (N, x_o) \to (P, y_o)$ is a stable map-germ and $\ell \geq p + 1$, then $A^{\ell}(N, P)$ has not only the above-mentioned property near $j^{\ell}f$, but still more is true; f has a representative $\tilde{f} : U \to V$ admitting a canonical stratification of which the stratification of the source coincides with $((J^{\ell}\tilde{f})^{-1} A^{\ell}(N, P))_{\tilde{f}}$. The last property clearly exhibits a certain universal character of $A^{\ell}(N, P)$. The proof of these statements will occupy the largest part of this section.

In the preceding chapter (III, 5.8) it has been shown that for any stable map-germ $f : (N, x_o) \to (P, y_o)$ we can find coordinate neighbourhoods $(U; x_1, \ldots, x_n)$ at (N, x_o), $(V; y_1, \ldots, y_p)$ at (P, y_o) and a representative $\tilde{f} : U \to V$ of f such that $y_j \circ \tilde{f}$ is a polynomial in $x_1, \ldots, x_n$ $(j = 1, \ldots, p)$. Let Σ denote the critical set of $\tilde{f}$. Let ϵ be so small that the closure of $B = \{x \in U: \ \Sigma_i \ x_i(x)^2 < \epsilon\}$ is contained in U. By taking ϵ even smaller, we may, following (1.5), also assume that $\tilde{f} | \Sigma \cap \bar{B}$ has finite fibres and that $\Sigma \cap \tilde{f}^{-1}(y_o) \cap B \subset \{x_o\}$. Next we choose $\eta > 0$ such that the closure of $D = \{y \in V: \ \Sigma_j \ y_j(y)^2 < \eta\}$ is contained in V and $\Sigma \cap \tilde{f}^{-1}(D) \cap \partial B = \phi$.

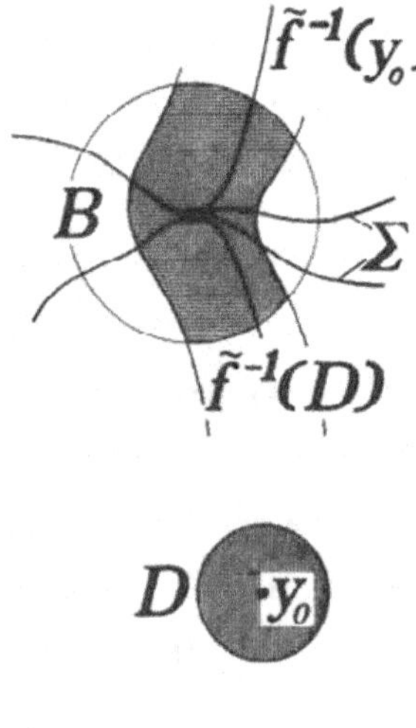

We claim that $\tilde{f}: \Sigma \cap \tilde{f}^{-1}(D) \cap B \to D$ is proper. Indeed, if $K \subset D$ is compact, then $\Sigma \cap \tilde{f}^{-1}(K) \cap B$ is a closed subset of the open relatively compact set $\tilde{f}^{-1}(D) \cap B$, and hence compact. We deduce that the restriction $\tilde{f}_o : \tilde{f}^{-1}(D) \cap B \to D$ of $\tilde{f}$ is a polynomial mapping between semialgebraic sets (with respect to the coordinates introduced above) which satisfies the Genericity Condition introduced in Chapter I. It has the additional property that $\tilde{f}_o^{-1}(y_o) / \{x_o\}$ contains no critical point. According to (I, 3.5) $\tilde{f}_o$ then admits a canonical Thom stratification $(\mathcal{A}, \mathcal{A}')$. In a sense the 'germ' of $(\mathcal{A}, \mathcal{A}')$ at (x_o, y_o) depends on f only. More precisely, if $\tilde{f}_1 : \tilde{f}_1^{-1}(D_1) \cap B_1 \to D_1$ is a representative of f obtained in the same way as $\tilde{f}_o$ and if $(\mathcal{B}, \mathcal{B}')$ denotes its canonical stratification, then it follows from (I, 3.8) and (I, 3.9) that $(\mathcal{A}, \mathcal{A}')$ and $(\mathcal{B}, \mathcal{B}')$ coincide on the pair $(\tilde{f}^{-1}(D \cap D_1) \cap B \cap B_1, D \cap D_1)$.

Since we are mainly interested in the properties of $(\mathcal{A}, \mathcal{A}')$ in a neighbourhood of (x_o, y_o) we would like any stratum of $\mathcal{A}$ to have x_o in its closure. The strata of $\mathcal{A}'$ will then have y_o in their closure. This can be realised as follows. Let V' be the subset of D obtained by removing the closed set $\bigcup\{\bar{X}' ;\ y_o \notin \bar{X}'$ and X' a connected component of a stratum of $\mathcal{A}'\}$. V' is an open subset of D, so by (I, 3.8) the pair $(\mathcal{A}|\tilde{f}^{-1}(V') \cap B, \mathcal{A}'|V')$ is the canonical stratification of the corresponding restriction of $\tilde{f}$. Now the stratification of the target is as required, so we turn our attention to the stratification of the source. We claim that any stratum X of $\mathcal{A}|\tilde{f}^{-1}(V') \cap B$ which does not contain x_o in its closure, necessarily avoids Σ: if not, then $X \subset \Sigma$, as $\Sigma \cap \tilde{f}^{-1}(D) \cap B$ is a union of strata. Because the restriction of $\tilde{f}$ to $\Sigma \cap \tilde{f}^{-1}(D) \cap B \to D$ is proper and $y_o \in \overline{\tilde{f}(X)}$, it follows that $\Sigma \cap \tilde{f}^{-1}(y_o) \cap \bar{X} \neq \phi$. Since $\tilde{f}^{-1}(y_o) \cap \bar{X} \subset \{x_o\}$, this implies $x_o \in \bar{X}$, whence a contradiction. It follows, that if U' is the union of connected components of strata of $\mathcal{A}|\tilde{f}^{-1}(V) \cap B$ with x_o in their closure, then $\Sigma \cap U' = \Sigma \cap \tilde{f}^{-1}(V) \cap B$. Moreover U' is open. By (I, 3.9) the pair $(\mathcal{A}|U', \mathcal{A}'|V')$ is then a canonical stratification of the restriction $\tilde{f}' : U' \to V'$ of f.

Representatives of f like $\tilde{f}'$ are sufficiently interesting to deserve a definition : a representative $\tilde{f}$ of a map-germ $f : (N, x_o) \to (P, y_o)$ is called special if (i) it satisfies the genericity condition, (ii) $\tilde{f}^{-1}(y_0) / \{x_0\}$ contains no critical point, and (iii) $\tilde{f}$ admits a canonical stratification such that x_0 lies in the closure of any connected component of a stratum in the source. For future reference we sum up:

(2.1) Any stable map-germ admits a special representative. The canonical stratifications of two such representatives coincide where both are defined.

Let f be a stable map-germ, $\tilde{f}$ a special representative of f and (A, A') the canonical stratification of $\tilde{f}$. It follows from (I, 1.1) that the stratum $X \in A$ which contains x_0 is connected and that its codimension is strictly greater than the codimension of any other stratum of A. By (2.1) this number only depends on f; we shall therefore call it the codimension of f and we write cod f. Clearly cod $f = o$ if and only if f is of maximal rank. This concept generalises to the case when f is only of finite singularity type. Then we let cod f be the codimension of some stable unfolding of f. To see that this is well-defined, let F and F' be stable unfoldings of f. By taking the cartesian product of one of them (say F) with the identity mapping id of some $(\mathbb{R}^k, o)$, we can arrange that the unfoldings $F \times \mathrm{id}$ and F' have the same unfolding dimension. Both $F \times \mathrm{id}$ and F' are stable and hence by (III, 4.4) smoothly equivalent. Then it follows from (I, 3.7) that cod $(F \times \mathrm{id})$ = cod F' and it is easy to see that cod F = cod $(F \times \mathrm{id})$ - see (I, 3.11).

Since contact-equivalent germs of finite singularity type have equivalent universal unfoldings by (III, 4.4), cod f only depends on the contact equivalence class of f. In particular, any jet which is contact-sufficient, has a well defined codimension. Our notion of codimension has of course little to do with, for instance, the notion of contact-codimension.

We recall from chapter III, that $W^\ell(N, P)$ stands for the set of $z \in J^\ell(N, P)$ with $\chi(z) \geq \ell$. Among other things it was shown there that $W^\ell(N, P)$ is a closed subset of $J^\ell(N, P)$ and that any $z \in J^\ell(N, P) / W^\ell(N, P)$ is contact-sufficient. We let $A^\ell(N, P)$ denote the partition of $J^\ell(N, P) / W^\ell(N, P)$

whose members are $S_j = \{z \in J^{\ell}(N, P) \setminus W^{\ell}(N, P) : \text{cod } z = j\}$, $j = 0, 1, 2, \ldots$. In the remainder of this section we investigate the local properties of this partition. In particular we will show that $A^{\ell}(N, P)$ is a Whitney stratification. For this purpose it is convenient to have the following lemma at our disposal.

(2.2) Lemma. *Let* M *be a manifold and* $\phi : M \to J^{\ell}(N, P)$ *a smooth mapping which is transverse to the contact classes. Let further* X *and* Y *be contact-invariant subsets of* $J^{\ell}(N, P)$.

(i) *If* $\phi^{-1}X$ *is a submanifold of* M, *then* X *is a submanifold of* $J^{\ell}(N, P)$ *of the same codimension near* $\phi(M) \cap X$.

(ii) *If* $\phi^{-1}X$ *and* $\phi^{-1}Y$ *are submanifolds of* M *and* $\phi^{-1}Y$ *is Whitney regular over* $\phi^{-1}X$, *then* Y *is Whitney regular over* X *near* $\phi(M) \cap X$.

Similar statements hold if contact class is replaced by diff $N \times$ diff P - orbit.

Proof. We only prove the contact-case; the other case can be treated in the same way. The natural projection $J^{\ell}(N, P) \to N \times P$ comes from a locally trivial fibre bundle with typical fibre $J^{\ell}(n, p)$ and structural group $L^{\ell}(n) \times L^{\ell}(p)$. Since the contact group ${}^{\ell}(n, p)$ contains $L^{\ell}(n) \times L^{\ell}(p)$, we may replace $J^{\ell}(N, P)$ by $J^{\ell}(n, p)$ and contact class by ${}^{\ell}(n, p)$-orbit, in the statements of the lemma, and prove (2.2) for this case.

Since $\phi : M \to J^{\ell}(n, p)$ is now transverse to ${}^{\ell}(n, p)$ - orbits, the mapping $\Phi : {}^{\ell}(n, p) \times M \to J^{\ell}(n, p)$, $\Phi(g, m) = g.\phi(m)$, must be a submersion. If X is a contact-invariant subset of $J^{\ell}(n, p)$, we have $\Phi^{-1}(X) = {}^{\ell}(n, p) \times \phi^{-1}X$. So if $\phi^{-1}X$ is smooth, then so is $\Phi^{-1}X$. As Φ is a submersion, this implies that X is smooth near $\phi(M) \cap X$. This proves (2.2.i). (2.2.ii) can be dealt with in a similar fashion.

The key result of this chapter is

(2.3) Proposition. *Let* $f : (N, x_o) \to (P, y_o)$ *be a stable map-germ and let* $\ell \geq \chi(f)$. *Then*

(a) $A^{\ell}(N, P)$ *is a Whitney stratification near* $j^{\ell}f$, *and* f *admits a special representative* $\tilde{f}$ *such that*

(b) $\tilde{f}$ is multi-transverse with respect to $A^{\ell}(N, P)$.

(c) if (A, A') is the canonical stratification of f, then A and $((J^{\ell}\tilde{f})^{-1} A^{\ell}(N, P))_{\tilde{f}}$ are equivalent, i.e. they have the same family of connected components of strata,

(d) the strata of A and $(J^{\ell}\tilde{f})^{-1} A^{\ell}(N, P)$ which contain x_o are equal.

We shall prove (2.3) in a number of steps. The proposition is trivially true if f is of maximal rank, so we may assume that $\operatorname{cod} f > o$. Following (2.1) f admits a special representative $\tilde{f} : U \to V$. Since $W^{\ell}(N, P)$ is closed and $j^{\ell}f \notin W^{\ell}(N, P)$, we may suppose that $J^{\ell}\tilde{f}(U) \cap W^{\ell}(N, P) \neq \emptyset$. By the transversality criterion (III, 6.11) we may also suppose that for any $x \in U$, $\tilde{f}_x$ is a stable map-germ. In the course of the proof below it is sometimes necessary to shrink U and V. If we do so, we tacitly assume that these conditions are still satisfied after shrinking. We let (A, A') denote the canonical stratification of $\tilde{f}$.

Step 1. Let $X \in A$ and $X^* \in (J^{\ell}\tilde{f})^{-1} A^{\ell}(N, P)$ contain x_o. Then the following two conditions are equivalent

(i) $x \in X$

(ii) $x \in X^*$ and $\tilde{f}^{-1}\tilde{f}(x) \cap \Sigma = \{x\}$.

Proof. The critical set Σ of $\tilde{f}$ is a union of strata of A. Hence so is $\Sigma \cap \tilde{f}^{-1}\tilde{f}(X)$. Since $\tilde{f}|\Sigma$ is proper and finite - to - one, $\tilde{f}|\Sigma \cap \tilde{f}^{-1}\tilde{f}(X)$ is also proper and finite - to - one. Because the strata in $\Sigma \cap \tilde{f}^{-1}\tilde{f}(X)$ are mapped submersively to $\tilde{f}(X)$, it follows that $\tilde{f} : \Sigma \cap \tilde{f}^{-1}\tilde{f}(X) \to \tilde{f}(X)$ is actually a covering map. Its fibre over $y_o \in \tilde{f}(X)$ consists of $\{x_o\}$. As X is connected, it follows that all its fibres consist of one point, in other words $\Sigma \cap \tilde{f}^{-1}\tilde{f}(x) = \{x\}$ for all $x \in X$. This proves part of (i) => (ii).

Conversely, let $x \in \Sigma$ have the property that $\Sigma \cap \tilde{f}^{-1}\tilde{f}(x) = \{x\}$. Then after removing strata of A and A' if necessary, we obtain a special representative of $\tilde{f}_x$. So if $Y \in A$ contains x, then $\operatorname{cod} \tilde{f}_x = \operatorname{co\,dim} Y$. In view of the preceding we may apply this to any $x \in X$. Then $X = Y$, so that

$\operatorname{cod} \tilde{f}_x = \operatorname{codim} X$. Since $\operatorname{codim} X = \operatorname{cod} f$, this implies $x \in X^*$. This completes the proof that (i) => (ii). But we can also apply this to an x as in (ii). Then we have $\operatorname{codim} Y = \operatorname{cod} \tilde{f}_x = \operatorname{cod} f$ (since $x \in X^*$) and hence Y and X have the same codimension. It follows that $Y = X$, in particular $x \in X$. This proves (ii) => (i).

Step 2. Let T_i denote the set of $z \in S_i$ which have a stable map-germ satisfying (2.3.d) for some special representative). Then T_i is a diff $N \times$ diff P - invariant submanifold of $J^\ell(N, P)$ of codimension i.

Proof. It is clear that T_i is diff $N \times$ diff P-invariant. To prove that it is smooth and of the correct codimension, let $z \in T_i$ and assume for convenience that $z = j^\ell f$. This allows us to stick to the notation introduced above. So we are given (after possibly shrinking U and V) $X = X^*$. Since $T_i \subset S_i$, we have $(J^\ell\tilde{f})^{-1} T_i \subset X^*$. Now, let $x \in X$. It follows from the implication (i) => (ii) of step 1, that $\tilde{f}$ is a special representative of $\tilde{f}_x$. Since $X = X^*$ and $\tilde{f}_x$ is a stable map-germ, this implies that $J^\ell\tilde{f}(x) \in T_i$. We deduce that $X \subset (J^\ell\tilde{f})^{-1}T_i$. Using $X = X^*$ again, we see that $X = (J^\ell\tilde{f})^{-1}T_i$. By (2.2.i) T_i is then a submanifold of $J^\ell(N, P)$ near $j^\ell f$. It is of the correct codimension, since $\operatorname{codim} T_i = \operatorname{codim} X = \operatorname{cod} f = i$.

The following step will be only used to prove the next one.

Step 3. Put $X_j^* = (J^\ell\tilde{f})^{-1}S_j$ and $X_j^{**} = (J^\ell\tilde{f})^{-1}T_j$ $(j = 0, 1, 2, \ldots)$. If $\cup\{X_j^* : j \geq \operatorname{cod} f\}$ $X \neq \emptyset$, then there exists a $y \in V\setminus\tilde{f}(X)$ such that $\tilde{f}^{-1}(y) \cap X_k^{**} \neq \emptyset$ for some $k \geq \operatorname{cod} f$.

Proof. Let Z_k denote the set of $y \in \tilde{f}(X_k^*)$ with the property that for any $x \in \tilde{f}^{-1}(y) \cap X_k^*$, X_k^* is contained in the stratum $X_x \in \mathcal{A}$ (with $x \in X_x$) near x. Clearly Z_k is open in $\tilde{f}(X_k^*)$. It is also dense in $\tilde{f}(X_k^*)$: if W is an open subset of $\tilde{f}(X_k^*)$, then a point of W, where $y \in W \mapsto \Sigma\{\operatorname{codim} X_x : x \in \tilde{f}^{-1}(y) \cap X_k^*\}$ takes its minimum necessarily belongs to Z_k.

We show that for $k > o$, Z_k is a submanifold of V of dimension $n - k$ and that $\tilde{f}^{-1}(Z_k) \cap X_k^* \subset X_k^{**}$. Let k be $> o$ and $x \in \tilde{f}^{-1}(Z_k) \cap X_k^*$. Using the fact that $\tilde{f}|\Sigma$ is finite-to-one and that A satisfies the frontier condition, it is easily seen that we can find a neighbourhood U_x of x in U such that $\tilde{f}^{-1}\tilde{f}(x') \cap \Sigma \cap U_x = \{x'\}$ for all $x' \in X_x \cap U_x$. Now, choose a special representative $\tilde{g}$ of $\tilde{f}_x$ defined on an open subset $U'_x \subset U_x$ such that $X_k^* \cap U'_x \subset X_x \cap U'_x$. Since $\tilde{f}^{-1}\tilde{f}(x') \cap \Sigma \cap U_x = \{x'\}$ for all $x' \in X_x \cap U'_x = X_k^* \cap U'_x$, it follows from Step 1 (applied to $\tilde{g}$) that $X_k^* \cap U'_x$ coincides with the stratum of the canonical stratification of $\tilde{g}$ which contains x. Hence $x \in X_k^{**}$. It also follows that $\tilde{f}(X_k^* \cap U'_x)$ (being a stratum of the canonical stratification of $\tilde{g}$) is a submanifold of V of the same dimension as X_k^{**}, namely $n - k$.

Now suppose that $\cup\{X_j^* : j \geq \operatorname{cod} f\} \setminus X \neq \emptyset$. We assume that j is the smallest number $\geq \operatorname{cod} f$ with this property. Then by step (1) (i) => (ii), $X_j^* \setminus X \neq \emptyset$. Since Z_j is an open-dense subset of $\tilde{f}(X_j^*)$ and $\tilde{f}(X)$ a closed subset of V, we then also have $Z_j \setminus \tilde{f}(X) \neq \emptyset$. Because Z_k is an $(n-k)$-manifold, it follows that

$$(Z_j \setminus \tilde{f}(X)) \setminus \bigcup_{k > j} \tilde{f}(X_k^*) \supset (Z_j \setminus \tilde{f}(X)) \setminus \bigcup_{k > j} \bar{Z}_k \neq \emptyset .$$

Since $X_k^* \setminus X = \emptyset$ for $\operatorname{cod} f \leq k < j$, any y in the left hand side of the inclusion above is as required.

We now put $i = \operatorname{cod} f$ and assume inductively that (2.3) holds for all stable map-germs of smaller codimension. So in particular $T_j = S_j$ near $J^\ell\tilde{f}(U)$ for $j < i$.

By (1.4) we may suppose that (after shrinking U and V if necessary) $\tilde{f}$ is multi-transverse with respect to each of the manifolds, T_j, $j \geq 0$.

Step 4. We have $X = X^*$, $J^\ell f \in T_i$ and A is equivalent to $((J^\ell\tilde{f})^{-1}\{S_o, \ldots, S_i\})_{\tilde{f}}$.

Proof. Let $y \in V \setminus \tilde{f}(X)$ be such that $\tilde{f}^{-1}(y) \cap \Sigma \subset U_j \setminus X_j^{**}$. We enumerate the distinct points of $\tilde{f}^{-1}(y) \cap \Sigma$ by $x_1, \ldots, x_r$ and choose a neighbourhood U_p of x_p such that $\tilde{f}|U_p : U_p \to \tilde{f}(U_p)$ is a special representative for

$\tilde{f}x_\rho$ $(\rho = 1, \ldots, r)$. Let (A_ρ, A'_ρ) denote the canonical stratification of $\tilde{f}|U_\rho$ and let $X_\rho \in A_\rho$ contain x_ρ. Further put $c_\rho = \operatorname{codim} X_\rho$ $(= \operatorname{cod} \tilde{f}x_\rho)$. By hypothesis, $x_\rho \in X^{**}_{c_\rho}$, so that (after possibly shrinking U_ρ) $X_\rho = X^{**}_{c_\rho} \cap U_\rho$. Since $\tilde{f}$ is multi-transverse with respect to $\{T_j\}_{j \geq 0}$, it follows that the manifold collection $\{X_1, \ldots, X_r\}$ has regular intersections relative $\tilde{f}$. Hence their images, $\tilde{f}(X_1), \ldots, \tilde{f}(X_r)$ are in general position in V. From the fact that $A'_1, \ldots, A'_r$ are Whitney stratifications, it is now easily deduced that $A'_1, \ldots, A'_r$ are in general position in an open neighbourhood V_y of y. Following (I, 3.10), $\cap A'_\rho | V_y$ then extends to the canonical stratification of $\tilde{f} : U_\rho\ (U_\rho \cap \tilde{f}^{-1}V_y) \to V_y$, and by (I, 3.9) it even extends to the canonical stratification (A_y, A'_y) of $\tilde{f} : \tilde{f}^{-1}V_y \to V_y$. Observe that the members of $\{X_1 \cap \tilde{f}^{-1}V_y, \ldots, X_r \cap \tilde{f}^{-1}V_y\}_{\tilde{f}}$ are strata of A_y. By (I, 3.8) we have $A_y = A | \tilde{f}^{-1}V_y$. Since $X_\rho \cap \tilde{f}^{-1}V_y = X^{**}_{c_\rho} \cap U_\rho \cap \tilde{f}^{-1}V_y$, it follows that the member of $\{X^{**}_0, X^{**}_1, X^{**}_2, \ldots\}_{\tilde{f}}$ which contains x_ρ coincides on $U_\rho \cap \tilde{f}^{-1}V_y$ with the corresponding stratum X_{x_ρ} of A. This in particular implies that $c_\rho = \operatorname{codim} X^{**}_{c_\rho} \leq \operatorname{codim} X_{x_\rho}$. As $X_{x_\rho} \not\subset X$ (recall that $y \notin \tilde{f}(X)$), we have $\operatorname{codim} X_{x_\rho} < i$, so it follows that $c_\rho < i$. We have now proved that the strata of A which intersect $\tilde{f}^{-1}(y) \cap \Sigma$ and the corresponding strata of $\{X^{**}_0, \ldots, X^{**}_{i=1}\}_{\tilde{f}} = ((J^\ell \tilde{f})^{-1}\{S_0, \ldots, S_{i-1}\})_{\tilde{f}}$ coincide (or are equivalent) on $\tilde{f}^{-1}V_y$. This proves at least a part of the last clause of step 4. It remains to show that $U_{j \geq i} X^*_j = X$: then clearly $X^*_i = X$ (for we know already from step 1 that $X^*_i = X^* \supset X$), which immediately implies $j^\ell f \in T_i$, while, in view of the preceding, this also completes the proof of the last clause. Now suppose that $U_{j \geq i} X^*{}_j \setminus X \neq \emptyset$. By step 3 there then exists a $y \in V \setminus \tilde{f}(X)$ such that $\tilde{f}^{-1}(y) \cap \Sigma \subset U_k X^{**}_k$ and $f^{-1}(y) \cap X^{**}_k \neq \emptyset$ for some $k \geq i$. But we have just seen that this cannot occur.

Step 5. For any j, S_j is Whitney regular over S_i at $j^\ell f$.

Proof. According to (2.2) it suffices to prove this for their counter images in U under $J^{\ell}\tilde{F}$. By step 4, $(J^{\ell}\tilde{F})^{-1} S_i = X$, while $(J^{\ell}\tilde{F})^{-1}S_j$ is a union of connected components of strata of A. Each of these is Whitney regular over X, hence so is their union $(J^{\ell}\tilde{F})^{-1}S_j$.

Step 5 implies (2.3-a), while it follows from step 4 that $\tilde{F}$ satisfies (2.3-b, c, d). The proof of (2.3) is now complete.

§3 Properties of the stratification

In this section we prove that $A^{\ell}(N, P)$ possesses the properties which were announced earlier. The main result here is (3.4), which states that under suitable transversality conditions (with respect to $A^{\ell}(N, P)$) a smooth family of mappings

$$N \times T \xrightarrow{F} P \times T \rightarrow T$$

(from N to P) admits a stratification satisfying almost all the hypotheses of Thom's Second Isotopy Lemma (II, 5.8). It is clear that this must bring us quite close to a proof of the topological stability theorem. But first we partially generalise (2.3).

(3.1) Proposition.

(a) $A^{\ell}(N, P)$ <u>is a Whitney stratification.</u>

Let $f : (N, x_0) \rightarrow (P, y_0)$ be a smooth map-germ such that $\chi(f) \leq \ell$, together with a stable unfolding

$$\begin{array}{ccc} (N, x_0) & \rightarrow & (P, y_0) \\ \downarrow i & & \downarrow j \\ (N', x_0') & \xrightarrow{F} & (P', y_0') \end{array}$$

and a special representative $\tilde{F} : U' \rightarrow V'$ of F. Denote by (A, A') the canonical stratification of $\tilde{F}$. Then

(b) i <u>is transverse to</u> A <u>if and only if</u> $J^{\ell}f$ <u>is transverse to</u> $A^{\ell}(N, P)$.

Moreover, if one of the conditions of (b) is satisfied, there exist representatives $\tilde{f} : U \to V$ of f, $\tilde{i} : U \to U'$ of i and $\tilde{j} : V \to V'$ of j such that

(c) $(\tilde{i}^{-1}A, \tilde{j}^{-1}A')$ is a Thom stratification of $\tilde{f}$.

(d) $\tilde{f}$ is multi-transverse with respect to $A^{\ell}(U, V)$ and $((J^{\ell}\tilde{f})^{-1}A^{\ell}(U, V))_{\tilde{f}}$ is equivalent to $\tilde{i}^{-1}A$.

Proof. Following (III, 0.1), by choosing suitable coordinates for (N', x_0') and (P', y_0'), we can write $(N', x_0') = (N \times T, x_0' \times t_0)$, $(P', y'_0) = (P \times T, y_0 \times t_0)$ such that i and j are the obvious embeddings and F is of the form $F(x, t) = (f_t(x), t)$ with $f_{t_0} = f$. Now it is clear that shrinking U' and V' (such that $\tilde{F}$ is still special) does not affect the statements of (3.1). Therefore, as the reader will verify without much trouble, we may assume that T, U' and V' are such that $U' = U \times T$ and $V' = V \times T$ (with U and V open neighbourhoods of $x_0 \in N$ and $y_0 \in P$ respectively) so that $\tilde{F}(x, t) = (\tilde{f}_t(x), t)$. Moreover by (III,6.11) we may suppose that the jet extension $J : U \times T \to J^{\ell}(N, P)$, $J(x, t) = J^{\ell}\tilde{f}_t(x)$, is transverse to all the contact classes in $J^{\ell}(N, P)$.

Now it is almost immediate from our definition of codimension that $J^{-1}A^{\ell}(N, P) = (J^{\ell}\tilde{F})^{-1} A^{\ell}(N \times T, P \times T)$. By (1.4) and (2.3-a) the latter is (after possibly shrinking $U \times T$ and $V \times T$) a Whitney stratification. It then follows from (2.2-ii) that $A^{\ell}(N, P)$ is a Whitney stratification near $j^{\ell}f$. This proves (a).

Let $X \in A$ contain (x_0, t_0). Following (2.3-d), X is also a stratum of $(J^{\ell}\tilde{F})^{-1} A^{\ell}(N \times Z, P \times Z) = J^{-1} A^{\ell}(N, P)$. So i is transverse to X if and only if $J \circ i$ is transverse to the stratum of $A^{\ell}(N, P)$ which contains $j^{\ell}f$. Since $J^{\ell}f = J \circ i$, this implies (b).

To prove the last two statements, we assume that $U \times \{t_0\}$ is transverse to A. Then $V \times \{t_0\}$ is clearly transverse to A'. So if we let $\tilde{i} : U \to U \times T$ and $\tilde{j} : V \to V \times T$ be the natural maps, and take $\tilde{f} = \tilde{F}_{t_0}$, then $(\tilde{i}^{-1}A, \tilde{j}^{-1}A')$ is a Thom stratification of $\tilde{f}$ indeed (easy to check, or consider this as a special case (I, 3.10)).

Finally, to prove that $\tilde{F}$ is multi-transverse with respect to $A^{\ell}(U, V)$ we would like to apply (1.6) with the following substitutions $M = U \times T$, $Q = V \times T$, $R = V \times \{z_o\}$ and $\mathcal{B} = (J^{\ell}\tilde{F})^{-1} A^{\ell}(N \times Z, P \times Z)$. Now it follows from (2.3-b, c) that the condition (1.6-i) is satisfied. Hence according to (1.6), $(J^{\ell}\tilde{F})^{-1} A^{\ell}(U, V)$ has regular intersections relative $\tilde{F}$. Clearly, $((J^{\ell}\tilde{F})^{-1}A^{\ell}(N, P))_{\tilde{F}}$ and $\tilde{F}^{-1}A$ are equivalent stratifications.

(3.2) Corollary. Let $f : (N, x_o) \to (P, y_o)$ be a smooth map-germ with an unfolding

$$\begin{array}{ccc} (N, x_o) & \xrightarrow{f} & (P, y_o) \\ \downarrow i & & \downarrow j \\ (N', x'_o) & \xrightarrow{F} & (P', y'_o) \end{array} .$$

Then the following two conditions are equivalent.

(i) $j^{\ell}f \notin W^{\ell}(N, P)$ and $J^{\ell}f$ is transverse to $A^{\ell}(N, P)$.

(ii) $j^{\ell}F \notin W^{\ell}(N', P')$, $J^{\ell}F$ is transverse to $A^{\ell}(N', P')$ and if $X \in (J^{\ell}F)^{-1}A^{\ell}(N', P')$ contains x'_o, then i is transverse to N'.

Proof. By (III, 2.5) we have $\chi(F) = \chi(f)$. Suppose that this number is finite. Then by (III, 2.8) F admits a stable unfolding G. If we assume that condition (i) is satisfied, then (ii) follows by applying (3.1-b) to the pair (f, G) first and then to the pair (F, G).

Proposition (3.1) directly leads to a stratification for certain mappings.

(3.3) Proposition. Let $f : N \to P$ be a proper smooth mapping which is multi-transverse with respect to $A^{\ell}(N, P)$ and such that $J^{\ell}f(N) \cap W^{\ell}(N, P) = \emptyset$. Put

$$A = ((J^{\ell}f)^{-1} A^{\ell}(N, P))_f \quad \text{and}$$
$$A' = \{f(X) : X \in A\} \cup \{P \setminus f(N)\} .$$

Then (A, A') Thom stratifies f.

Proof. Let $y \in P$. Since the critical set Σ is a union of strata of $(J^{\ell}f)^{-1} A^{\ell}(N, P)$ each of dimension $< p$, the multi-transversality property implies that $f^{-1}(y) \cap \Sigma$ contains at most p points. Let $x_1, \ldots, x_r$ be the distinct points of $f^{-1}(y) \cap \Sigma$. Following (3.1 - c, d), f_{x_ρ} admits a representative $g_\rho : U_\rho \to V_\rho$ with a Thom stratification (A_ρ, A'_ρ) such that $A_\rho = ((J^\ell g_\rho)^{-1} A^\ell(N, P))_{g_\rho}$. Though it was not explicitly mentioned in (3.1), it is clear that $A'_\rho = \{g_\rho(X) : X \in A_\rho\} \cup \{V_\rho \setminus g_\rho(U_\rho)\}$. We may moreover assume that if $X_\rho \in A_\rho$ contains x_ρ, then X_ρ is also a stratum of $(J^\ell g_\rho)^{-1} A^\ell(N, P)$. By multi-transversality, their images $g_1(X_1), \ldots, g(X_r)$ are in regular position at y. Because these are strata of the Whitney stratifications $A'_1, \ldots, A'_r$, it is easily shown that then $A'_1, \ldots, A'_r$ must be in general position in an open neighbourhood V of y in $V_1 \cap \ldots \cap V_r$. Following (I, 3.10), the intersection $(A'_1 \cap \ldots \cap A'_r) | V$ then naturally extends to a Thom stratification of $f : (U_1 \cup \ldots \cup U_r) \cap f^{-1}V \to V$. It is easy to see that this stratification is just the restriction of (A, A'). It follows that $(A|f^{-1}V, A'|V)$ is a stratification of $f : f^{-1}V \to V$.

The previous proposition generalises to families of such mappings. This has the interesting corollary (3.5) below.

(3.4) Proposition. <u>Let</u> T <u>be a smooth manifold and let</u> $F : N \times T \to P \times T$ <u>be a smooth mapping of the form</u> $F(x, t) = (f_t(x), t)$. <u>Then the following two conditions are equivalent.</u>

(i) <u>For all</u> $t \in T$, $J^\ell f_t(N) \cap W^\ell(N, P) = \emptyset$ <u>and</u> f_t <u>is multi-transverse with respect to</u> $A^\ell(N, P)$.

(ii) $J^\ell F(N \times T) \cap W^\ell(N \times T, P \times T) = \emptyset$, F <u>is multi-transverse with respect to</u> $A^\ell(N \times T, P \times T)$ <u>and the restriction of the projection</u> $\pi_T : N \times T \to T$ <u>to any stratum of</u> $((J^\ell F)^{-1} A^\ell(N \times T, P \times T))_F$ <u>is a submersion.</u>

In either case we have that for all t

$$((J^\ell F)^{-1} A^\ell(N \times T, P \times T))_F \big| N \times \{t\} = ((J^\ell f_t)^{-1} A^\ell(N, P))_{f_t} \times \{t\}.$$

Proof. Assume that condition (i) is satisfied. It then follows from (3.2) that $J^\ell F$ avoids $W^\ell(N \times T, P \times T)$ and that $J^\ell F$ is transverse to $A^\ell(N \times T, P \times T)$. Now, define $J : N \times T \to J^\ell(N, P)$ by $J(x, t) = J^\ell f_t(x)$.

Then $J^{-1}A^{\ell}(N, P) = (J^{\ell}F)^{-1}A^{\ell}(N \times T, P \times T)$. Since $J|N \times \{t\}$ is transverse to $A^{\ell}(N, P)$, $N \times \{t\}$ is transverse to $J^{-1}A^{\ell}(N, P)$. As this is so for all $t \in T$, it follows that π_T maps the strata of $J^{-1}A^{\ell}(N, P)$ submersively to T. We are also given that $J^{-1}A^{\ell}(N, P)|N \times \{t\}$ has regular intersections relative $F : N \times \{t\} \to P \times \{t\}$, so that we may apply (1.7, ii => i). It follows that $J^{-1}A^{\ell}(N, P) = (J^{\ell}F)^{-1}A^{\ell}(N \times T, P \times T)$ has regular intersections relative F and that the strata of $((J^{\ell}F)^{-1}A^{\ell}(N \times T, P \times T))_F$ are transverse to $N \times \{t\}$ for all $t \in T$. This proves (ii) as well as the last statement. The proof that (ii) implies (i) uses (1.7, i => ii) and is left to the reader.

A little more than the hypotheses of (3.4) is needed to ensure that the family $\{f_t\}_{t \in T}$ is topologically trivial.

(3.5) Corollary (to (3.3), (3.4) and (I, 5.8)) *Keep the notations of (3.4) and assume that one of the conditions* (3.4 - i, ii) *is satisfied. Suppose moreover that* F *is proper and that there exists a proper smooth function* $\phi : P \to \mathbb{R}$ *such that the restriction of the composite* $\phi \circ f_t : N \to \mathbb{R}$ *to any stratum of* $((J^{\ell}f_t)^{-1}A^{\ell}(N, P))_{f_t}$ *is transverse to* $\mathbb{Z} \subset \mathbb{R}$. *Then* F *defines a locally trivial family of mappings, i.e. for any* $t \in T$, *there exist a neighbourhood* U *of* $t \in T$ *and homeomorphisms* $h : N \times U \to N \times U$, $h' : P \times U \to P \times U$ *such that the diagram below commutes*

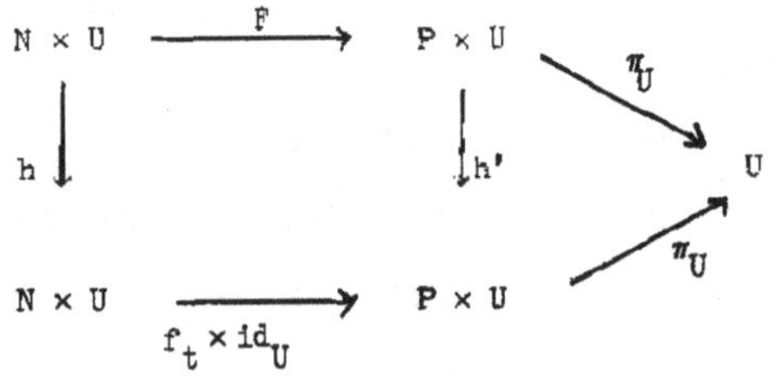

In particular, any $f_{t'}$ $(t' \in U)$ *is topologically equivalent to* f_t.

Observe that if P is compact such a ϕ always exists; we simply let $\phi(P) = \{\frac{1}{2}\}$. The existence of ϕ is postulated in order to have some control on F at infinity in case P is not compact.

<u>Proof</u> Let (A_t, A'_t) and (A, A') denote the Thom stratifications of $f_{t'}$ and F respectively, obtained by applying (3.3). Then by (3.4) A is transverse to $N \times \{t'\}$ and $A|N \times \{t'\}$ corresponds with A_t. Our aim is to let an innocent refinement of (A, A') satisfy the hypotheses of the Second Isotopy Lemma. The assumptions regarding ϕ imply that $(\phi \circ \Pi_P \circ F)^{-1}\mathbb{Z}$ is transverse to each intersection $X \cap (N \times \{t'\}$; $X \in A$, $t' \in T$. Hence $\Pi_T \circ F$ maps any intersection $X \cap (\phi \circ \Pi_P \circ F)^{-1}\mathbb{Z}$, $X \in A$, submersively to T. Now, if we let $\mathcal{C}$ denote the stratification of $\mathbb{R}$ whose strata are the connected components of $\mathbb{Z}$ and $\mathbb{R} - \mathbb{Z}$, then A and $(\phi \circ \Pi_T \circ F)^{-1}\mathcal{C}$ are in regular position and hence so are A' and $(\phi \circ \Pi_T)^{-1}\mathcal{C}$. We put $B = A \cap (\phi \circ \Pi_T \circ F)^{-1}\mathcal{C}$ and $B' = A' \cap (\phi \circ \Pi_P)^{-1}\mathcal{C}$. The pair (B, B') Thom stratifies F and since $\phi_T \circ F$ maps the strata of B submersively to T, Π_T, $\phi \circ \Pi_T$ does the same with the strata of B'. Moreover each stratum of B' is contained in a subset $\phi^{-1}[n, n+1] \times T$. Since $\phi^{-1}[n, n+1]$ is compact, the restriction of Π_T to the closure of any stratum of B' is proper. As F is proper, the restriction of F to the closure of any stratum of B is also proper. Hence the hypotheses of (II, 5.8) are satisfied for the diagram

$$(N \times T, B) \xrightarrow{F} (P \times T, B') \xrightarrow{\Pi_T} T .$$

The result follows from (II,5.8) and (II,5.9).

§4 Topological stability as a transversality property

We now aim to prove that (for ℓ sufficiently large) any proper smooth mapping $f : N \to P$ which is multi-transverse with respect to $A^{\ell}(N, P)$ is topologically stable. In view of (1, 1) this will imply the topological stability theorem. The progress we have made so far in this direction is best summarised by (3.5). With an application of this last result in mind the following proposition must be crucial.

(4.1) Proposition. The proper smooth mappings $f : N \to P$ which are multi-transverse with respect to $A^{\ell}(N, P)$ and satisfy $J^{\ell}f(N) \cap W^{\ell}(N, P) = \emptyset$ form an open subset $\Omega^{\ell}(N, P)$ of $C^{\infty}(N, P)$.

The proof requires a bit of preparation in the form of lemmas (4.2) and (4.3) below.

(4.2) Lemma. Let $f \in C^{\infty}(N, P)$ and let $\{f_j \in C^{\infty}(N, P)\}_{j=1}^{\infty}$ converge to f. Then a subsequence of $\{f_j\}$ embeds in a one-parameter family, more precisely, there exists a family $\{F_t : N \to P\}_{t \in \mathbb{R}}$ of mappings such that $(x, t) \mapsto F_t(x)$ is smooth and $\{F_{1/k}\}_{k=1}^{\infty}$ is a subsequence of $\{f_j\}_{j=1}^{\infty}$. In particular, $F_0 = f$.

Proof. Choose a Riemannian metric for N such that N is complete with respect to this metric and do the same for P. These determine for every ℓ a metric ρ_{ℓ} on $J^{\ell}(N, P)$ for which $J^{\ell}(N, P)$ is complete.

We choose a tube in the sense of (II, 1.4, 1.5) for the diagonal of $P \times P$ as a submanifold of $P \times P$, compatible with the projection onto the first factor. This exists by (II, 1.6). The tube furnishes us a neighbourhood V_P of the diagonal together with a smooth mapping $\gamma : V_P \times [0, 1] \to P$ such that $\gamma(y_1, y_2, 0) = y_1$ and $\gamma(y_1, y_2, 1) = y_2$.

We now select a subsequence $\{f_{j(k)}\}_{k=1}^{\infty}$ of $\{f_j\}_{j=1}^{\infty}$ such that

$\sup\{\rho_{\ell}(J^{\ell}f_{j(k)}(x), J^{\ell}f(x)) : x \in N\} < k^{-k^2}$ and $(f_{j(k+1)}(x), f_{j(k)}(x)) \in V_P$

for all $x \in N$. Let $\sigma : \mathbb{R} \to [0, 1]$ be a smooth function which vanishes on $(-\infty, 0]$ and equals 1 on $[1, \infty)$. Then define $g : N \times \mathbb{R} \to P$ by
$g(x, t) = \gamma(f_{j(k+1)}(x), f_{j(k)}(x), \sigma(t + 1 - k))$ if $t \in [k - 1, k]$ and

$g(x, t) = f(x)$ if $t < 0$. Then g is a smooth C^{∞}-mapping on $N \times (\mathbb{R} \setminus \{0\})$ with the property that $\sup\{\rho_{\ell}(J^{\ell}g_t(x), J^{\ell}f(x)) : x \in N\} < k^{-k^2}$ if $|t| > k$. We leave it to the reader to deduce that the mapping $(x, t) \mapsto g(x, t^{-1})$ on $N \times (\mathbb{R} \setminus \{0\})$ extends smoothly over $N \times \mathbb{R}$ by giving it the value $f(x)$ in $(x, 0)$. It follows that the family $\{F_t\}_{t \in \mathbb{R}}$ defined by $F_t(x) = g(x, t^{-1})$ $(t \neq 0)$ and $F_0(x) = f(x)$ is as required.

We use (4.2) to prove the following

(4.3) Lemma. <u>Let</u> $f \in \Omega^{\ell}(N, P)$ <u>and let</u> K <u>be a compact subset of</u> P. <u>Then the set</u> Ω_K <u>of</u> $f' \in C^{\infty}(N, P)$ <u>with</u> $f'|f^{-1}K$ <u>multi-transverse with respect to</u> $A^{\ell}(N, P)$ <u>is a neighbourhood of</u> f <u>in</u> $C^{\infty}(N, P)$.

Proof. Suppose not. As f has a countable neighbourhood basis there then exists a sequence $\{f_j \in C^{\infty}(N, P)\}_{j=1}^{\infty}$ converging to f such that $f_j|f^{-1}K$ is not multi-transverse with respect to $A^{\ell}(N, P)$. (As usual, such a condition always refers to a neighbourhood of $f^{-1}K$ in N.) By the previous lemma (4.2) there exists a smooth one-parameter family $\{F_t : N \to \mathbb{R}\}_{t \in \mathbb{R}}$ with $F_o = f$ and $\{F_{1/k}\}_{k=1}^{\infty}$ a subsequence of $\{f_j\}_{j=1}^{\infty}$. It follows from (3.2) that there exists an $\epsilon > 0$ such that $J^{\ell}F$ is transverse to $A^{\ell}(N \times \mathbb{R}, P \times \mathbb{R})$ at $K \times [-\epsilon, \epsilon]$. Following (3.2), $N \times \{0\}$ is transverse to $(J^{\ell}F)^{-1}A^{\ell}(N \times \mathbb{R}, P \times \mathbb{R})$ and their intersection coincides with $(J^{\ell}f)^{-1}A^{\ell}(N, P) \times \{0\}$. Since $(J^{\ell}f)^{-1}A^{\ell}(N, P)$ has regular intersections relative f, it follows from (1.7) that for a possibly smaller positive ϵ, $(J^{\ell}F)^{-1}A^{\ell}(N \times \mathbb{R}, P \times \mathbb{R}) \mid F^{-1}(K \times [-\epsilon, \epsilon])$ has regular intersections relative F, and that the strata of

$$A_{K,\epsilon} = (J^{\ell}F)^{-1}A^{\ell}(N \times \mathbb{R}, P \times \mathbb{R}) \mid F^{-1}(K \times [-\epsilon, \epsilon])_F$$

are transverse to $N \times \{0\}$. Since $A_{K,\epsilon}$ is a Whitney stratification by (3.3), we may then choose ϵ so small that $N \times \{t\}$ is transverse to $A_{K,\epsilon}$ for all $t \in [-\epsilon, \epsilon]$. Then (1.7) implies that $F_t|F_t^{-1}K$ is multi-transverse with respect to $A^{\ell}(N, P)$ for all $t \in [-\epsilon, \epsilon]$. In particular this is so for $t = \frac{1}{k}$ if $k > \epsilon^{-1}$, which contradicts our assumption.

Proof of (4.1). Let $f \in \Omega^{\ell}(N, P)$. Choose a pair $\{V_i \subset W_i\}_{i \in I}$ of locally finite coverings of P by relatively compact open subsets for which $\bar{V}_i \subset W_i$ for all $i \in I$. The set Ω of $f' \in C^{\infty}(N, P)$ with $J^{\ell}f'(N) \cap W^{\ell}(N, P) = \emptyset$ and $f'(f^{-1}\bar{V}_i) \subset W_i$ for all $i \in I$ is clearly open. By the last condition any $f' \in \Omega$ is proper. On the other hand it follows from (4.3) that $\bigcap_{i \in I} \Omega_{\bar{V}_i}$ ($\Omega_{\bar{V}_i}$ as there defined) is a neighbourhood of f. Hence their intersection $\Omega \cap \bigcap_{i \in I} \Omega_{\bar{V}_i}$ is a neighbourhood of f. It is clear that this intersection is contained in $\Omega^{\ell}(N, P)$.

(4.4) Theorem. <u>Any</u> $f \in \Omega^{\ell}(N, P)$ <u>is topologically stable</u>.

(4.5) Corollary. (<u>The Topological Stability Theorem</u>) <u>The topologically stable mappings intersect the set of proper smooth mappings</u> $C^{\infty}_{pr}(N, P)$ <u>in a dense subset.</u>

Proof. In view of (4.4) it suffices to show that $\Omega^{\ell}(N, P)$ is dense in $C^{\infty}_{pr}(N, P)$ for some ℓ. By (III, 7.2) there exists an ℓ such that codim $W^{\ell}(n, p) > n$. The density of $\Omega^{\ell}(N, P)$ in $C^{\infty}_{pr}(N, P)$ will follow from (1.1) once we have show that $W^{\ell}(N, P)$ admits a stratification into strata of codimension $> n$. This is easily done as follows. Recall that $W^{\ell}(n, p)$ is a semi-algebraic subset of $J^{\ell}(n, p)$ (III, 7.1). Hence it admits by (I, 2.7) a canonical stratification. This stratification is then invariant under the structural group $L^{\ell}(n) \times L^{\ell}(p)$ of the jet-bundle $J^{\ell}(N, P) \to N \times P$ and so determines a stratification of $W^{\ell}(N, P)$ into strata of codimension $> n$.

Proof of (4.4). By choosing a tube for the diagonal in $P \times P$ we obtain (as in the proof of (4.2)) a neighbourhood V_P of the diagonal in $P \times P$ and a smooth mapping $\gamma : V_P \times [0,1] \to P$ such that $\gamma(y_1, y_2, 0) = y_1$ and $\gamma(y_1, y_2, 1) = y_2$. Now let $f \in \Omega^{\ell}(N, P)$. Then (3.3) constructs out of $(J^{\ell}f)^{-1}A^{\ell}(N, P)$ a stratification (A, A') for f. Let $\phi : P \to \mathbb{R}$ be a proper smooth function such that the restriction of ϕ to any stratum of A' is transverse to $\mathbb{Z} \subset \mathbb{R}$. For any $j \in \mathbb{Z}$ we put $P_j = \phi^{-1}\{j\}$, $N_j = (\phi \circ f)^{-1}\{j\}$ and if $f' \in C^{\infty}(N, P)$ we set $N^{f'}_j = (\phi \circ f')^{-1}\{j\}$ and let $f'_j : N^{f'}_j \to P_j$ denote the restriction of f'. Note that N_j and P_j are <u>compact</u> submanifolds. It follows from (3.4) that $f_j \in \Omega^{\ell}(N_j, P_j)$. If f' is close to f, then $N^{f'}_j$ is close to N_j, so if π is a smooth retraction of a neighbourhood of N_j onto N_j, then $\pi|N^{f'}_j$ will be a diffeomorphism. This enables us to compare f'_j with f_j. Then (4.1) applied to f and f_j $(j \in \mathbb{Z})$, yields a neighbourhood Ω_f of f such that $f' \in \Omega_f$ implies

(a) For any $t \in [0, 1]$, the mapping $f'_t \in C^\infty(N, P)$, defined by $f'_t(x) = \gamma(f(x), f'(x), t)$, belongs to $\Omega^\ell(N, P)$.

(b) The composite $\phi \circ f'_t$ is transverse to $\mathbb{Z} \subset \mathbb{R}$ for any $t \in [0, 1]$ and $f'_{t,j} \in \Omega^\ell(N_j^{f'_t}, P_j)$ for any $j \in \mathbb{Z}$. We prove that any $f' \in \Omega_f$ is topologically equivalent to f. According to (3.3) property (a) implies that $(J^\ell f'_t)^{-1} A^\ell(N, P)$ determines a stratification (A_t, A'_t) of f'_t. From (3.4) and property (b) above it follows that the strata of A'_t are transverse to the manifolds P_j, $j \in \mathbb{Z}$, in particular, that the restriction of ϕ f_t to any stratum of A_t is transverse to $\mathbb{Z}$. It then follows from (3.5) that f and f' are topologically equivalent.

□□

References

[Bo] Borel, A. Linear Algebraic Groups, W.A. Benjamin, Inc. New York, 1969.

[Br] Bröcker, T. & Jänich, K. Einführung in die Differential topologie. Springer - Verlag, 1973.

[Gi] Gibson, C.G. Regularity of the Segre Stratification, Math. Proc. Camb. Phil. Soc. (To appear)

[Go] Golubitsky, M. & Guillemin, V. Stable Mappings and their Singularities, Springer - Verlag, 1973.

[Hi] Hironaka, H. "Number Theory, algebraic geometry and commutative algebra". Volume in honour of Y. Akizuki. Published by Kinokuniya, Tokyo, 1973.

[La] Lang, S. Differential Manifolds, Addison - Wesley, 1972.

[Ło] Łojasiewicz, S. Ensembles Semi-Analytiques, IHES Lecture Notes, 1965.

[Lo] Looijenga, E.J.N. Structural Stability of Smooth Families of C^{∞} - functions. Thesis, University of Amsterdam, 1974.

[Mo] Morlet, C. Seminaire H. Cartan, Exposé 4, 1961 - 62.

[Ma] Mather, J. N. I. Notes on Topological Stability, Lecture Notes, Harvard University, 1970.

II. Stratifications and Mappings, Proceedings of the Dynamical Systems Conference, Salvador, Brazil, July, 1971, Academic Press.

III. Finitely-determined map germs. Publ. Math. IHES 35 (1968) pp.127 - 156.

IV. Classification of Stable Germs by $\mathbb{R}$ - algebras. Publ. Math. IHES 37 (1969) pp.223 - 248.

V. Transversality. Advances in Mathematics 4 (1970) pp.301 - 336.

VI. The Nice Dimensions. Proc. Liverpool Singularities Symposium I, Springer Lecture Notes in Maths. 192 (1971).

[Th] Thom, R. Propriétés Différentielles Locales des Ensembles Analytiques, Seminaire Bourbaki, 1964/5. exp. 281.

[Wa] Wall, C.T.C. Regular Stratifications. "Dynamical Systems - Warwick 1974". Springer Lecture Notes in Math. No. 468, p.332 - 344.

[Wa] Wasserman, G. Stability of Unfoldings, Springer Lecture Notes in Mathematics, 393 (1974).

[Wh] Whitney, H. I. Local Properties of Analytic Varieties, pp.205 - 244 in Differential and Combinatorial Topology, Princeton, 1965.

II. Tangents to an Analytic Variety, Annals. of Maths. 81 (1965), 496 - 549.

Index

Symbols

Vol. 399: Functional Analysis and its Applications. Proceedings 1973. Edited by H. G. Garnir, K. R. Unni and J. H. Williamson. II, 584 pages 1974.

Vol. 400: A Crash Course on Kleinian Groups. Proceedings 1974. Edited by L. Bers and I. Kra. VII, 130 pages. 1974.

Vol. 401: M F Atiyah, Elliptic Operators and Compact Groups. V, 93 pages. 1974

Vol. 402: M Waldschmidt, Nombres Transcendants VIII, 277 pages. 1974.

Vol. 403: Combinatorial Mathematics Proceedings 1972. Edited by D. A Holton. VIII, 148 pages 1974

Vol. 404: Théorie du Potentiel et Analyse Harmonique Edité par J. Faraut. V, 245 pages. 1974.

Vol. 405: K. J. Devlin and H Johnsbråten, The Souslin Problem. VIII, 132 pages 1974

Vol. 406: Graphs and Combinatorics Proceedings 1973 Edited by R. A. Bari and F. Harary. VIII, 355 pages 1974

Vol 407: P Berthelot, Cohomologie Cristalline des Schémas de Caracteristique p > o. II, 604 pages. 1974.

Vol. 408: J Wermer, Potential Theory VIII, 146 pages 1974

Vol 409: Fonctions de Plusieurs Variables Complexes, Séminaire François Norguet 1970–1973. XIII, 612 pages. 1974

Vol. 410: Séminaire Pierre Lelong (Analyse) Année 1972–1973 VI, 181 pages 1974

Vol. 411: Hypergraph Seminar. Ohio State University, 1972. Edited by C. Berge and D. Ray-Chaudhuri. IX, 287 pages 1974.

Vol 412: Classification of Algebraic Varieties and Compact Complex Manifolds. Proceedings 1974 Edited by H Popp V, 333 pages 1974.

Vol. 413: M Bruneau, Variation Totale d'une Fonction XIV, 332 pages. 1974.

Vol 414: T. Kambayashi, M Miyanishi and M Takeuchi, Unipotent Algebraic Groups VI, 165 pages 1974

Vol. 415: Ordinary and Partial Differential Equations Proceedings 1974. XVII, 447 pages. 1974.

Vol. 416: M E Taylor, Pseudo Differential Operators IV, 155 pages 1974

Vol. 417: H. H. Keller, Differential Calculus in Locally Convex Spaces XVI, 131 pages 1974

Vol 418: Localization in Group Theory and Homotopy Theory and Related Topics Battelle Seattle 1974 Seminar Edited by P J Hilton VI, 172 pages 1974.

Vol. 419: Topics in Analysis. Proceedings 1970. Edited by O. E. Lehto, I. S Louhivaara, and R. H. Nevanlinna. XIII, 392 pages. 1974.

Vol. 420: Category Seminar. Proceedings 1972/73. Edited by G M. Kelly. VI, 375 pages. 1974.

Vol 421: V Poénaru, Groupes Discrets VI, 216 pages 1974

Vol. 422: J.-M Lemaire, Algèbres Connexes et Homologie des Espaces de Lacets XIV, 133 pages 1974

Vol. 423: S. S. Abhyankar and A. M. Sathaye, Geometric Theory of Algebraic Space Curves. XIV, 302 pages. 1974.

Vol. 424: L Weiss and J Wolfowitz, Maximum Probability Estimators and Related Topics V, 106 pages 1974

Vol 425: P R Chernoff and J E Marsden, Properties of Infinite Dimensional Hamiltonian Systems IV, 160 pages 1974

Vol. 426: M. L. Silverstein, Symmetric Markov Processes. X, 287 pages. 1974

Vol 427: H Omori, Infinite Dimensional Lie Transformation Groups XII, 149 pages 1974

Vol 428: Algebraic and Geometrical Methods in Topology, Proceedings 1973 Edited by L F McAuley. XI, 280 pages 1974.

Vol. 429: L Cohn, Analytic Theory of the Harish-Chandra C-Function III, 154 pages. 1974.

Vol. 430: Constructive and Computational Methods for Differential and Integral Equations. Proceedings 1974. Edited by D L. Colton and R. P. Gilbert VII, 476 pages. 1974

Vol 431: Séminaire Bourbaki – vol. 1973/74. Exposés 436–452. IV, 347 pages 1975.

Vol. 432: R P. Pflug, Holomorphiegebiete, pseudokonvexe Gebiete und das Levi-Problem. VI, 210 Seiten. 1975.

Vol. 433: W. G. Faris, Self-Adjoint Operators. VII, 115 pages. 1975.

Vol 434: P. Brenner, V. Thomée, and L. B. Wahlbin, Besov Spaces and Applications to Difference Methods for Initial Value Problems. II, 154 pages. 1975.

Vol. 435: C. F. Dunkl and D. E. Ramirez, Representations of Commutative Semitopological Semigroups. VI, 181 pages. 1975

Vol. 436: L. Auslander and R. Tolimieri, Abelian Harmonic Analysis, Theta Functions and Function Algebras on a Nilmanifold. V, 99 pages 1975.

Vol. 437: D. W. Masser, Elliptic Functions and Transcendence. XIV, 143 pages 1975.

Vol 438: Geometric Topology Proceedings 1974. Edited by L C Glaser and T B Rushing. X, 459 pages 1975

Vol 439: K. Ueno, Classification Theory of Algebraic Varieties and Compact Complex Spaces. XIX, 278 pages. 1975

Vol 440: R. K Getoor, Markov Processes: Ray Processes and Right Processes. V, 118 pages 1975

Vol. 441: N. Jacobson, PI-Algebras An Introduction. V, 115 pages. 1975

Vol. 442: C. H. Wilcox, Scattering Theory for the d'Alembert Equation in Exterior Domains. III, 184 pages. 1975.

Vol 443: M. Lazard, Commutative Formal Groups. II, 236 pages. 1975

Vol. 444: F. van Oystaeyen, Prime Spectra in Non-Commutative Algebra. V, 128 pages. 1975.

Vol 445: Model Theory and Topoi. Edited by F. W. Lawvere, C Maurer, and G. C. Wraith. III, 354 pages. 1975.

Vol 446: Partial Differential Equations and Related Topics. Proceedings 1974. Edited by J A. Goldstein. IV, 389 pages. 1975

Vol 447: S. Toledo, Tableau Systems for First Order Number Theory and Certain Higher Order Theories III, 339 pages 1975.

Vol. 448: Spectral Theory and Differential Equations. Proceedings 1974. Edited by W. N. Everitt. XII, 321 pages. 1975.

Vol 449: Hyperfunctions and Theoretical Physics Proceedings 1973 Edited by F. Pham IV, 218 pages. 1975

Vol. 450: Algebra and Logic. Proceedings 1974. Edited by J. N. Crossley VIII, 307 pages. 1975.

Vol. 451: Probabilistic Methods in Differential Equations Proceedings 1974 Edited by M A. Pinsky VII, 162 pages 1975.

Vol 452: Combinatorial Mathematics III Proceedings 1974. Edited by Anne Penfold Street and W D Wallis. IX, 233 pages. 1975

Vol. 453: Logic Colloquium. Symposium on Logic Held at Boston, 1972–73. Edited by R. Parikh. IV, 251 pages. 1975.

Vol. 454: J. Hirschfeld and W. H Wheeler, Forcing, Arithmetic, Division Rings. VII, 266 pages. 1975.

Vol. 455: H. Kraft, Kommutative algebraische Gruppen und Ringe. III, 163 Seiten 1975

Vol. 456: R. M. Fossum, P. A. Griffith, and I. Reiten, Trivial Extensions of Abelian Categories Homological Algebra of Trivial Extensions of Abelian Categories with Applications to Ring Theory XI, 122 pages. 1975.

Vol. 457: Fractional Calculus and Its Applications Proceedings 1974. Edited by B. Ross. VI, 381 pages. 1975.

Vol 458: P Walters, Ergodic Theory – Introductory Lectures VI, 198 pages 1975

Vol. 459: Fourier Integral Operators and Partial Differential Equations Proceedings 1974. Edited by J Chazarain. VI, 372 pages. 1975

Vol. 460: O. Loos, Jordan Pairs. XVI, 218 pages. 1975

Vol. 461: Computational Mechanics. Proceedings 1974. Edited by J. T. Oden. VII, 328 pages. 1975.

Vol. 462: P. Gérardin, Construction de Séries Discrètes p-adiques. »Sur les séries discrètes non ramifiées des groupes réductifs déployés p-adiques«. III, 180 pages 1975.

Vol. 463: H.-H. Kuo, Gaussian Measures in Banach Spaces. VI, 224 pages. 1975.

Vol. 464: C Rockland, Hypoellipticity and Eigenvalue Asymptotics. III, 171 pages 1975

Vol. 465: Séminaire de Probabilités IX Proceedings 1973/74 Edité par P A Meyer. IV, 589 pages 1975.

Vol. 466: Non-Commutative Harmonic Analysis. Proceedings 1974 Edited by J Carmona, J Dixmier and M. Vergne. VI, 231 pages. 1975

Vol. 467: M. R Essén, The Cos $\pi\lambda$ Theorem. With a paper by Christer Borell VII, 112 pages 1975.

Vol. 468: Dynamical Systems – Warwick 1974 Proceedings 1973/74 Edited by A. Manning. X, 405 pages. 1975.

Vol. 469: E. Binz, Continuous Convergence on C(X). IX, 140 pages 1975.

Vol. 470: R. Bowen, Equilibrium States and the Ergodic Theory of Anosov Diffeomorphisms. III, 108 pages. 1975.

Vol. 471: R. S. Hamilton, Harmonic Maps of Manifolds with Boundary III, 168 pages. 1975.

Vol. 472: Probability-Winter School. Proceedings 1975. Edited by Z. Ciesielski, K. Urbanik, and W. A. Woyczyński VI, 283 pages. 1975.

Vol. 473: D. Burghelea, R. Lashof, and M. Rothenberg, Groups of Automorphisms of Manifolds. (with an appendix by E. Pedersen) VII, 156 pages. 1975.

Vol. 474: Séminaire Pierre Lelong (Analyse) Année 1973/74. Edité par P. Lelong. VI, 182 pages. 1975.

Vol 475: Répartition Modulo 1 Actes du Colloque de Marseille-Luminy, 4 au 7 Juin 1974. Edité par G. Rauzy V, 258 pages. 1975 1975.

Vol. 476: Modular Functions of One Variable IV. Proceedings 1972. Edited by B J. Birch and W. Kuyk V, 151 pages 1975.

Vol. 477: Optimization and Optimal Control. Proceedings 1974. Edited by R. Bulirsch, W. Oettli, and J. Stoer. VII, 294 pages 1975.

Vol. 478: G Schober, Univalent Functions – Selected Topics. V, 200 pages. 1975.

Vol. 479: S. D Fisher and J. W. Jerome, Minimum Norm Extremals in Function Spaces With Applications to Classical and Modern Analysis VIII, 209 pages 1975.

Vol. 480: X M Fernique, J P Conze et J Gani, Ecole d'Eté de Probabilités de Saint-Flour IV–1974 Edité par P.-L. Hennequin. XI, 293 pages. 1975.

Vol. 481: M. de Guzmán, Differentiation of Integrals in R^n. XII, 226 pages 1975.

Vol. 482: Fonctions de Plusieurs Variables Complexes II. Séminaire François Norguet 1974–1975 IX, 367 pages. 1975.

Vol. 483: R. D. M. Accola, Riemann Surfaces, Theta Functions, and Abelian Automorphisms Groups. III, 105 pages. 1975.

Vol. 484: Differential Topology and Geometry. Proceedings 1974. Edited by G. P. Joubert, R. P. Moussu, and R. H Roussarie IX, 287 pages 1975.

Vol 485: J. Diestel, Geometry of Banach Spaces – Selected Topics. XI, 282 pages 1975.

Vol. 486: S. Stratila and D Voiculescu, Representations of AF-Algebras and of the Group U (∞) IX, 169 pages. 1975

Vol. 487: H. M. Reimann und T. Rychener, Funktionen beschränkter mittlerer Oszillation. VI, 141 Seiten 1975.

Vol 488: Representations of Algebras, Ottawa 1974 Proceedings 1974 Edited by V. Dlab and P. Gabriel XII, 378 pages 1975.

Vol 489: J. Bair and R Fourneau, Etude Géométrique des Espaces Vectoriels. Une Introduction. VII, 185 pages. 1975.

Vol 490: The Geometry of Metric and Linear Spaces Proceedings 1974 Edited by L M. Kelly X, 244 pages 1975

Vol. 491: K A Broughan, Invariants for Real-Generated Uniform Topological and Algebraic Categories. X, 197 pages. 1975

Vol. 492: Infinitary Logic: In Memoriam Carol Karp. Edited by D. W. Kueker. VI, 206 pages. 1975.

Vol. 493: F. W. Kamber and P. Tondeur, Foliated Bundles and Characteristic Classes. XIII, 208 pages 1975.

Vol. 494: A Cornea and G. Licea Order and Potential Resolvent Families of Kernels. IV, 154 pages 1975.

Vol. 495: A Kerber, Representations of Permutation Groups II V, 175 pages. 1975

Vol. 496: L H. Hodgkin and V P. Snaith, Topics in K-Theory Two Independent Contributions III, 294 pages. 1975

Vol 497: Analyse Harmonique sur les Groupes de Lie. Proceedings 1973–75. Edité par P Eymard et al. VI, 710 pages. 1975

Vol. 498: Model Theory and Algebra. A Memorial Tribute to Abraham Robinson. Edited by D. H. Saracino and V. B. Weispfenning. X, 463 pages 1975.

Vol. 499: Logic Conference, Kiel 1974. Proceedings. Edited by G. H. Müller, A. Oberschelp, and K Potthoff. V, 651 pages 1975.

Vol 500: Proof Theory Symposion, Kiel 1974. Proceedings. Edited by J Diller and G H. Müller VIII, 383 pages. 1975.

Vol. 501: Spline Functions, Karlsruhe 1975. Proceedings. Edited by K. Böhmer, G. Meinardus, and W. Schempp. VI, 421 pages. 1976.

Vol. 502: János Galambos, Representations of Real Numbers by Infinite Series. VI, 146 pages 1976.

Vol 503: Applications of Methods of Functional Analysis to Problems in Mechanics. Proceedings 1975. Edited by P. Germain and B. Nayroles. XIX, 531 pages 1976

Vol 504: S Lang and H F Trotter, Frobenius Distributions in GL_2-Extensions III, 274 pages 1976

Vol 505: Advances in Complex Function Theory. Proceedings 1973/74 Edited by W. E Kirwan and L Zalcman. VIII, 203 pages 1976

Vol. 506: Numerical Analysis, Dundee 1975 Proceedings Edited by G A. Watson. X, 201 pages 1976.

Vol 507: M C. Reed, Abstract Non-Linear Wave Equations VI, 128 pages 1976

Vol 508: E Seneta, Regularly Varying Functions. V, 112 pages. 1976.

Vol. 509: D E. Blair, Contact Manifolds in Riemannian Geometry. VI, 146 pages 1976

Vol 510: V. Poènaru, Singularités C^∞ en Présence de Symétrie V, 174 pages. 1976

Vol 511: Séminaire de Probabilités X Proceedings 1974/75. Edité par P A. Meyer VI, 593 pages 1976

Vol 512: Spaces of Analytic Functions, Kristiansand, Norway 1975. Proceedings. Edited by O. B Bekken, B. K. Øksendal, and A. Stray. VIII, 204 pages. 1976.

Vol. 513: R. B. Warfield, Jr. Nilpotent Groups. VIII, 115 pages. 1976.

Vol. 514: Séminaire Bourbaki vol. 1974/75. Exposés 453 – 470. IV, 276 pages 1976.

Vol 515: Bäcklund Transformations Nashville, Tennessee 1974 Proceedings. Edited by R. M Miura. VIII, 295 pages. 1976.